W9-BVT-441

Probably Approximately Correct

Probably Approximately Correct

Nature's Algorithms for Learning and Prospering in a Complex World

LESLIE VALIANT

BASIC BOOKS
A Member of the Perseus Books Group
New York

Published by Basic Books,
A Member of the Perseus Books Group

A CIP catalog record for this book is available from the Library of Congress.
ISBN: 978-0-465-03271-6
ISBN (e-book): 978-0-465-03790-2
10 9 8 7 6 5 4 3 2 1

Contents

Summary

Algorithms are the step-by-step instructions used in computing for achieving desired results, much like recipes in cooking. In both cases the recipe designer has a certain controlled environment in mind for realizing the recipe, and foresees how the desired outcome will be achieved. The algorithms I discuss in this book are special. Unlike most algorithms, they can be run in environments unknown to the designer, and they learn by interacting with the environment how to act effectively in it. After sufficient interaction they will have expertise not provided by the designer, but extracted from the environment. I call these algorithms ecorithms. The model of learning they follow, known as the probably approximately correct model, provides a quantitative framework in which designers can evaluate the expertise achieved and the cost of achieving it.

These ecorithms are not merely a feature of computers. I argue in this book that such learning mechanisms impose and determine the character of life on Earth. The course of evolution is shaped entirely by organisms interacting with and adapting to their environments. This biological inheritance, as well as further learning from the environment after conception and birth, have a determining influence on the course of an individual's life. The focus here will be the unified study of the mechanisms of evolution, learning, and intelligence using the methods of computer science.

The book has the following simple structure. Chapters 1, 2, and 4 set the scene for the natural phenomena to which the quantitative computational approach is to be applied. Chapter 3 is an introduction to computer science, particularly the quantitative study of algorithms and their complexity, and describes the background for the methodology used. Chapters 5, 6, and 7 contain the resulting theory for learning, evolution, and intelligence, respectively.

The final chapters make some informal and more speculative suggestions with regard to some consequences for humans and machines.

Mathematics

The language of mathematics will be used, but only a little, and will be explained where used.

Chapter One

Ecorithms

In 1947 John von Neumann, the famously gifted mathematician, was keynote speaker at the first annual meeting of the Association for Computing Machinery. In his address he said that future computers would get along with just a dozen instruction types, a number known to be adequate for expressing all of mathematics. He went on to say that one need not be surprised at this small number, since 1,000 words were known to be adequate for most situations in real life, and mathematics was only a small part of life, and a very simple part at that. The audience reacted with hilarity. This provoked von Neumann to respond: "If people do not believe that mathematics is simple, it is only because they do not realize how complicated life is."[1]

Though counterintuitive, von Neumann's quip contains an obvious truth. Einstein's theory of general relativity is simple in the sense that one can write the essential content on one line as a single equation. Understanding its meaning, derivation, and consequences requires more extensive study and effort. However, this formal simplicity is striking and powerful. The power comes from the implied generality, that knowledge of one equation alone will allow one to make accurate predictions about a host of situations not even conceived when the equation was first written down.

Most aspects of life are not so simple. If you want to succeed in a job interview, or in making an investment, or in choosing a life partner, you can be quite sure that there is no equation that will guarantee you success. In these endeavors it will not be possible to limit the pieces of knowledge that might be relevant to any one definable source. And even if you had all the relevant knowledge, there may be no surefire way of combining it to yield the best decision.

This book is predicated on taking this distinction seriously. Those aspects of knowledge for which there is a good predictive theory, typically a mathematical or scientific one, will be called *theoryful*. The rest will be called *theoryless*. I use the term theory here in the same sense as it is used in science, to denote a "good, effective, and useful theory" rather than the negative sense of "only a theory." Predicting the orbit of a planet based on Newton's laws is theoryful, since the predictor uses an explicit model that can accurately predict everything about orbits. A card player is equally theoryful in predicting an opponent's hand, if this is done using a principled calculation of probabilities, as is a chemist who uses the principles of chemistry to predict the outcome of mixing two chemicals.

In contrast, the vast majority of human behaviors look theoryless. Nevertheless, these behaviors are often highly effective. *These abundant theoryless but effective behaviors still lack a scientific account, and it is these that this book addresses.*

The notions of the theoryful and the theoryless as used here are relative, relative to the knowledge of the decision maker in question. While gravity and mechanics may be theoryful to a physicist, they will not be to a fish or a bird, which still have to cope with the physical world, but do so, we presume, without following a theory. Worms can burrow through the ground without apparently any understanding of the physical laws to which they are subject. Most humans manage their finances adequately in an economic world they don't fully understand. They can often muddle through even at times when experts stumble. Humans can also competently navigate social situations that are quite complex, without being able to articulate how.

In each of these examples the entity manages to *cope* somehow, without having the tenets of a theory or a scientific law to follow. Almost any biological or human behavior may be viewed as some such coping. Many instances of effective coping have aspects both of the mundane and also of the grand and mysterious. In each case the behavior is highly effective, yet if we try to spell out exactly how the behavior operates, or why it is successful, we are often stumped. How can such behavior be effective in a world that is too complex to offer a clear scientific theory to be followed as a guide? Even more puzzling, *how can a capability for such effective coping be acquired in the first place?*

Science books generally restrict their subject matter to the theoryful. However, I am impressed with how effectively life forms "cope" with the theoryless in this complex world. Surely these many forms of coping have some commonality. Perhaps behind them all is a single basic phenomenon that is itself subject to scientific laws.

This book is based on two central tenets. The first is that the coping mechanisms with which life abounds are all the result of learning from the environment. The second is that this learning is done by concrete mechanisms that can be understood by the methods of computer science.

On the surface, any connection between coping and computation may seem jarring. Computers have traditionally been most effective when they follow a predictive science, such as the physics of fluid flow. However, computers also have their softer side. Contrary to common perception, computer science has always been more about humans than about machines. The many things that computers can do, such as search the Web, correct our spelling, solve mathematical equations, play chess, or translate from one language to another, all emulate capabilities that humans possess and have some interest in exercising. Depending on the task, the performance of present-day computers will be better or worse than humans. But in regarding computers merely as our slaves for getting things done, we may be missing the point. The overlap between what computers and humans do every day is already vast and diverse. Even without any extrapolation into the future, we have to ask what computers already teach us about ourselves.

The variety of applications of computation to domains of human interest is a totally unexpected discovery of the last century. There is no trace of anyone a hundred years ago having anticipated it. It is a truly awesome phenomenon. Each of us can identify our own different way of being impacted by the range of applications that computers now offer. A few years ago I was interested in the capabilities of a certain model of the brain. In a short, hermit-like span of a few weeks I ran a simulation of this model on my laptop and wrote up a paper based on the calculations performed by my laptop. I used a word processor on the same laptop to write and edit the article. I then emailed it off to a journal again from that laptop. This may sound unremarkable to the present-day reader, but a few generations ago, who would have thought that one device could perform such a variety of tasks? Indeed, while for most ideas some long and complex history can be traced, the modern notion of computation emerged remarkably suddenly, and in a most complete form, in a single paper published by Alan Turing in 1936.[2]

Science prior to that time made no mention of abstract machines. Turing's theory did. He defined the mathematical notion of computation that our all-pervasive information technology now follows. But in offering his work, he made it clear that his goal went beyond understanding only machines: "We may compare a man in the process of computing a real number to a machine which is only capable of a finite number of conditions." With these words he was declaring that he was aiming to formalize the process of computation where a human mechanically follows some rules. He was seeking to capture the limits of what could be regarded as mechanical intellectual work, where no appeal to other capabilities such as intuition or creativity was being made.

Turing succeeded so well that the word computation is now used in exactly the sense in which he defined it. We forget that a "computer" in the 1930s referred to a human being who made a living doing routine calculations. Speculations that philosophers or psychologists entertained in earlier times as to the nature of mechanical mental capabilities equally dim in the memory. Turing had discovered a precise and fundamental law that both living and inert things must obey, but which only humans had been observed to exhibit up to that time. His notion is now being realized in billions of pieces of technology that have transformed our lives. But if we are blinded by this technological success, we may miss the more important point that Turing's concept may enable us to understand human activity itself.

This may seem paradoxical. Humans clearly existed before Turing, but Turing's notion of computation was not noticed before his time. So how can his theory be so fundamental to humans if little trace of it had even been suspected before?

My answer to this is that even in the pre-Turing era, in fact since the beginning of life, the dominating force on Earth within all its life forms *was* computation. But the computations were of a very special kind. These computations were weak in almost every respect when compared with the capabilities of our laptops. They were exceedingly good, however, at one enterprise: adaptation. These are the computations that I call ecorithms—algorithms that derive their power by learning from whatever environment they inhabit, so as to be able to behave effectively in it. To understand these we need to understand computations in the Turing sense. But we also need to refine his definitions to capture the more particular phenomena of learning, adaptation, and evolution.

Understanding learning has been one of my personal research goals for several decades. The natural phenomenon of young children learning is extraordinary. A spectacular facet of this learning is that, beyond remembering individual experiences, children will also generalize from those experiences, and very quickly. After seeing a few examples of apples or chairs, they know how to categorize new examples. Different children see different examples, yet their notions become similar. When asked to categorize examples they have not seen before, their rate of agreement will be remarkably high, at least within any one culture. Young children can sort apples from balls even when both are round and red.

This ability to generalize looks miraculous. Of course, it cannot really be a miracle. It is a highly reproducible natural phenomenon. Ripe apples fall from the tree to the ground predictably enough that one can base a universal law of gravitation on this phenomenon. Children generalizing successfully from their specific experiences manifest a similarly predictable phenomenon, which therefore also begs for a scientific explanation. I seek to explain this in terms of concrete computational processes.

The phenomenon of generalization has been widely discussed by philosophers for millennia. It has been called the problem of induction. I have found that as a scientist I have some advantages over philosophers: It is sufficient to aim to capture the fundamental part of a specific reproducible phenomenon. I need not explain all of the many senses in which the words induction or generalization have been used. Scientific discovery—for example, Johannes Kepler discovering his laws of planetary orbits—may have some commonality with the phenomenon of generalization exhibited by children learning words, but it may be a secondary and harder to reproduce by-product of a more basic and fundamental capability. Turing did not attempt to capture all the connotations that the word computing may have had in his day. He sought only to uncover a phenomenon associated with that word that had fundamental reality independent of any word usage.

What kind of explanation of induction do we need? Does it need to be mathematical? There is no better answer to this than what is implicit in the work of Turing himself. I have already referred to his successful mathematical formulation of computation. But he is also famous for the notion that is now known as the Turing Test, which he offered as a test for recognizing whether a machine can be considered to think. A simplified definition is as follows. A machine passes the Turing Test if a person, conversing with it via remote electronic interactions, cannot distinguish it from a person. The Turing Test is an important notion, and researchers in artificial intelligence have not succeeded in either building machines that can pass the test or in showing it to be irrelevant. However, it is an informal notion. Unlike Turing's mathematical definition of computation, it does not tell us how exactly to proceed in order to emulate thinking. As a result, it has not led to progress in artificial intelligence remotely comparable to the success of general computation.

Hence the right goal must be to find a *mathematical* definition of learning of a nature similar to Turing's notion of computation, rather than an *informal* notion like the Turing Test. After all, where would we be if Turing had given for computation only an informal definition? Let us think about that. What would have been an informal notion of the "mechanically computable" that would have sounded plausible in Turing's time? How about this: "A task is mechanically computable if and only if it can be computed by a person of average intelligence while at the same time doing a mundane but exacting task, such as eating spaghetti." Few could have disputed the *reasonableness* of such a definition. But I doubt such a definition in 1936 could have spawned the twenty-first century we see around us.

At the heart of my thesis here is a mathematical definition of learning. It is called the PAC or the probably approximately correct model of learning, and its main features are the following:[3] The learning process is carried out by a concrete computation that takes a limited number of steps. Organisms cannot spend so long computing that they have no time for anything else or die before they finish. Also, the computation requires only a similarly limited number of interactions with the world during learning. Learning should enable organisms to categorize new information with at most a small error rate. Also, the definition has to acknowledge that induction is not logically fail-safe: If the world suddenly changes, then one should not expect or require good generalization into the future.

The biology of living organisms can be described in terms of complex circuits or networks that act within and between cells. Our biology is based on proteins and the interactions among them. Our DNA contains more than 20,000 genes that describe various proteins. Additionally, the DNA encodes descriptions of the regulation mechanism, a specification of how much new protein of each kind is to be produced, or expressed. This overall regulation mechanism is absolutely fundamental to our biology, and is called the protein expression network. It is of enormous complexity. Even though many of its details remain to be discovered, we can ask: *How have these well-functioning, highly intricate networks with so many interlocking parts come into being?* I believe that all these circuits are the result of some learning process instigated by the interactions between a biological entity and its environment.

Life's interactions can be viewed in terms of either a single organism's lifetime or the longer spans during which genes and species evolve. In either case the information gained by the entity from the interaction is processed in some mechanical way by what I call an ecorithm. The primary purpose of the ecorithm is to change the circuits so that they will behave better in the environment in the future and produce a better outcome for the owner.

Human biochemistry is an important enough topic. However, our neural circuits, comprising some tens of billions of neurons, may be viewed as being involved in our personal experiences even more intimately. Our psychological behavior is controlled by these circuits. How do these circuits arise in evolution, and how are they updated during life? By the same arguments they too must be the result of information obtained from interactions, by ourselves or our ancestors, and incorporated in our genes or brain by some adaptive mechanism.

If biological circuits are fundamentally shaped by learning processes, there seems little chance of understanding them, or their manifestations in our psychology, unless we recognize their origins in learning. We may not yet know in detail the actual ecorithms used in biology on Earth. However, the fact that our behaviors have their origins in such learning algorithms already has implications.

Earlier I listed as two central tenets that the behaviors that need explanation all arose from learning, and that this learning can be understood as a computational process. These tenets are not offered here as mere unproved assumptions, but as the consequences of the assumption that life has a mechanistic explanation.

The argument that these tenets actually follow from the formulation of ecorithms goes as follows: I start with the mechanistic assumption that biological forms came into existence as a result of concrete mechanisms operating in some environments. These mechanisms have been of two kinds, those that operate in individuals interacting with their environment, and those that operate via genetic changes over many generations.

I then make two observations. First, ecorithms are defined broadly enough that they encompass any mechanistic process. This follows from the work of Turing and his contemporaries that established the principle, known as the Church-Turing Hypothesis, that all processes that can be regarded as mechanistic can be captured by a single notion of computation or algorithm. Second, ecorithms are also construed broadly enough to encompass any process of interaction with an environment. From these two observations one can conclude that the coping mechanisms of nature have no sources of influence on them that are not fully accounted for by ecorithms, simply because we have defined ecorithms broadly enough to account for all such influences.

To put this in a different way, the news reported here is that there is a burgeoning science of learning algorithms. Once the existence of such a science is accepted, its centrality to the study of life is more or less self-evident.

Of course, the reader should be cautious when confronted with purported logical arguments such as the one I just gave. Indeed, later chapters will address the general pitfalls of reasoning about theoryless subject matter. It is appropriate, therefore, to attempt to corroborate my proposition. Is there somewhere we can turn for a sanity check? The answer is machine learning, a method for finding patterns in data that are usefully predictive of future events but which do not necessarily provide an explanatory theory.

Machine learning is already a widely used technology with diverse applications. For example, companies such as Amazon and Netflix make recommendations to shoppers based on the predictions of learning algorithms trained on past data. Of course, there is no theory of which books or movies you will like. You may even completely change your tastes at any time. Nevertheless, using machine learning algorithms, it is possible to do a useful job in making such recommendations. Financial institutions likewise use machine learning algorithms, in their case, for example, for detecting whether individual credit card purchase attempts are likely to be fraudulent. These algorithms pick up various kinds of relevant information, such as the geographical pattern of your previous purchases, to make some decisions based on data collected from many past transactions. The development of the learning algorithms used may well be theoryful. But this again does not mean that fraud itself is theoryful. New kinds of fraud are being invented all the time. The algorithms merely find patterns in past credit card purchases that are useful enough to give financial institutions a statistical edge in coping with this area of the theoryless.

Much of everyday human decision making appears to be of a similar nature—it is based on a competent ability to predict from past observations without any good articulation of how the prediction is made or any claim of fundamental understanding of the phenomenon in question. The predictions need not be perfect or the best possible. They need merely to be useful enough. The fact that these algorithms are already in widespread use, and produce useful results in areas most would regard as theoryless, is good evidence that we are on the right track.

However, the idea of an ecorithm goes well beyond the idea of machine learning in its current, general usage. Within the study of ecorithms several additional notions beyond the learning algorithms themselves are included. First, there is the notion that it is important to specify what we expect a learning algorithm to be able to do before we can declare it to be successful. Second, using such a specification, we can then discuss problems that are not learnable—some environments will be so complex that it is impossible for any entity to cope. Third, there is the question of how broad a functionality one wants to have beyond generalization in the machine learning sense. To have intelligent behavior, for example, one needs at least a reasoning capability on top of learning. Finally, biological evolution must fit somehow into the study of coping mechanisms, but it is not clear exactly how, since traditional views of evolution do not exactly fit the machine learning paradigm. In studying ecorithms, we want to embrace all of these issues, and more.

The problem of dealing with the theoryless is ever present in our lives. Every day we are forced to put our trust in the judgment of experts who operate outside the bounds of any strict science. Your doctor and car mechanic are paid to make judgments, based on their own experience and that of their teachers. We presume that their expertise is the result of learning from a substantial amount of real-world experience and, for that reason, is effective in coping with this complex world. Their expertise can be evaluated by how well their diagnoses and predictions work out. In some areas we can evaluate performance, at least after the fact.

We are also exposed every day to commentators and pundits whose diagnoses and predictions are infrequently checked for ultimate accuracy. We hear about what will happen in politics, the stock market, or the economy, but these predictions often seem hardly better than random guessing.

In late 2008 Queen Elizabeth II asked a group of academics why the world financial crisis had not been predicted. She was not the only one asking this question. Was the crisis inherently unpredictable in some sense, or was the failure due to some gross negligence? After the crisis a substantial amount of public discussion pertained to this question. Is there a rational way of predicting rare events? Why do humans have so many intellectual frailties and behave as irrationally as they do? Why are humans subject so easily to deception and self-deception? Why do humans systematically delude themselves into thinking that they are good predictors of future events even if they are not?

Many reasons have been given for the difficulty of making predictions, and the mistakes that people are prone to make have been widely analyzed.[4] The following, for example, is an instructive argument. After any significant historical event numerous explanations of the causes are offered. These explanations can be so beguilingly plausible that we easily mistake them for actual causes that might have been detected before the events in question. We are then communally led into the belief that world events have identifiable causes and are generally predictable. Hence popular disappointment that the world financial crisis had not been better anticipated can be ascribed to widespread overexpectation and naïveté with regard to the possibility of making predictions.

This book departs from this approach and takes an opposing, more positive view. While making predictions may be inherently difficult, and we humans have our special failings, human predictive abilities are substantial and reason enough for some celebration. Humans, and biological systems generally, do have an impressive capability to make predictions. The ability of living organisms to survive each day in this dangerous world is surely evidence of an ability to predict the consequences of their actions and those of others, and to be prepared for whatever happens, and be rarely taken totally by surprise. In human terms, the fact that we can go through a typical day, one that may include many events and interactions with others, and be seldom surprised is testament surely of our good predictive talents. Of course, the domains in which we make these reliable predictions often relate only to everyday life—what other people will say or other drivers do. They are mundane, almost by definition. But even mundane predictions become mystifying once one tries to understand the process by which the predictions are being made, or tries to reproduce them in a computer.

From this viewpoint, the general disappointment that the world financial crisis had not been better predicted was not based entirely on naïve illusion. It was based on the well-justified high regard we have for our predictive abilities, and so it would be clearly to our advantage to identify why they failed. It may be that the world was changing in such a random fashion that the past did not even implicitly contain reliable information about the future. Or perhaps the past did indeed contain this information, but that it was somehow so complex that it was not practically feasible to dig it out. A third case is that prediction was indeed feasible, but the wrong algorithm or the wrong data had been used.

The study of ecorithms is concerned with delineating among these possibilities. Having the ability to make these distinctions among topics of everyday concern, such as predictions about the course of the economy, seems important. One may be able to do more than merely lament human frailties in this regard. Are there inherent reasons why reliable predictions are not possible regarding the course of a country's economy? Perhaps one can show that there are. It would then follow that there is no reason to listen to pundits other than for entertainment.

Computation allows one to construct concrete situations in which the world does reveal sufficient information for prediction in principle, but not in practice. Consider the area of encryption. If messages in the wireless connection of your home computer are encrypted, the intention is that if your neighbor listens in, he should not be able to get any information about what you are doing. Even if he listens in over a long period and does clever computations on the data he collects using a powerful computer, he should not be able to invade your privacy. This is another way of saying that the environment defined by your enciphered messages should be too complex for your neighbor, or anyone else, to make any sense of.

How can entities cope with what they do not fully understand? The simplest living organisms have had to face this problem from the beginnings of life. With limited mechanisms they had to survive in a complex world and to reproduce. Every evolving species has faced a similar problem, as do individual humans going through their daily lives. I shall argue that solutions to these problems have to be sought within the framework of learning algorithms, since this is the mechanism by which life extracts information from its environment. By the end of the book I hope to have persuaded the reader that when seeking to understand the fundamental character of life, learning algorithms are a good place to start.

Chapter Two

Prediction and Adaptation

Only adapt.
ADAPTED FROM E. M. FORSTER

"You never walk into a situation and believe that you know better than the natives. You have to listen and look around. Otherwise you can make some very serious mistakes."[1] This was a lesson that Kofi Annan, the former Secretary General of the United Nations learned, not on some far-flung diplomatic posting for the UN, but as a young man in St. Paul, Minnesota. He had arrived from Africa to study economics as an undergraduate. Inexperienced as he was with cold weather, when he first saw local students wearing ear muffs he thought they looked ridiculous. But after walking round the campus on a cold day, he went out to buy some for himself.

The logic of ecorithms has much in common with Annan's analysis. That logic emphasizes listening and looking around. It encourages caution in applying specialized expertise gained in one environment to another, and gives respectful deference to observed experience. It says that it is we who must seek to adapt.

Such an adaptive imperative is absent from most aphorisms. "Neither a borrower nor a lender be" urges one to act in a specific way rather than to adapt to one's environment. The pitfalls of following such nonadaptive advice are clear. While the advice may be good in some circumstances, perhaps those from which it was derived, in others it may not be.

Annan's strategy has the strength that it accepts that there are many possible worlds and warns against assuming that they are all the same. On the other hand, it is not too specific in prescribing a course of action. I shall argue that some of the most important phenomena of biology and cognition arise from general adaptive strategies akin to Annan's, empty as they may appear to be of any specific expert knowledge. Although such strategies as listening and looking are not fine-tuned to any particular environment, they may nonetheless be effective in any environment that has certain weak regularities hidden among all the complexities. I shall suggest that not only are they effective, but, further, they are integral to any explanation of life and culture as we witness these on Earth.

The new word ecorithm that I use to encapsulate these ideas derives from the word *algorithm* and the prefix *eco-*. An algorithm is simply any well-defined procedure. It is derived from the Latinized transliteration Algoritmi of the name of the mathematician Al-Khwārizmī, who worked in the House of Wisdom in Baghdad in the ninth century and authored an influential book on algebra. I invoke the word algorithm intentionally. In the domain in which it is most widely used, namely computer science, the standards of explicitness—of what is considered well defined—are high. In the words of computer scientist Donald Knuth, "Science is what we understand well enough to explain to a computer. Art is everything else we do."[2] I want to discuss evolution, learning, and intelligence in terms of algorithms that are unambiguous and explicit enough that they can be "explained to," and hence simulated by, a computer. The prefix *eco-*, from the ancient Greek word *oikos* meaning household or home (and which evokes the word ecology), signals that we are interested in algorithms that operate in complicated environments, especially environments that are much more complex than the algorithm itself. There is no contradiction in this. While the algorithm has to perform well in a complex environment, about which it has little knowledge initially, it has a chance of doing so if it is allowed to interact extensively with the environment and learn from it.

Within the realm of computation I make the following distinction. Algorithms as traditionally studied in mathematics and computer science are designed to solve instances of particular problems, such as solving algebraic equations or searching for a word in a text. All the expertise they need for their success is encoded in their own description by their designer. For example, Euclid in his textbook *The Elements* describes an elegant algorithm for finding the greatest common divisor of two numbers. (The greatest common divisor of 30 and 42 is 6.) His algorithm is correct and efficient in a specifiable sense even for arbitrarily large numbers. Its exact behavior on all pairs of numbers is entirely predictable, and no doubt foreseen by Euclid.

Ecorithms are special algorithms. In contrast with those designed to solve specific mathematical problems, these operate in environments that are not fully known to the designer, and may have much arbitrariness. Nevertheless, ecorithms can perform well even in these environments. While their success is foreseeable, the actual course they take will vary according to the environment.

The requirements that such an algorithm must meet to offer a plausible explanation of a natural phenomenon, such as biological evolution, are quite onerous. In particular, the algorithm must achieve its goals after a limited number of interactions and with the expenditure of limited resources. The concept of ecorithms and the general model of learning in which they are embedded, which I call probably approximately correct (or PAC) learning, insist on such quantitative practicality. The phenomena that they seek to explain are some of the most familiar to human experience: learning, resilience, and adaptation. I argue that broader phenomena still, in particular evolution and intelligence, are also best understood in these terms.

Evolution in biology is the idea that life forms have changed over time, and that these changes have resulted in the organisms seen on Earth today. Although closely associated with Charles Darwin, the roots of the idea reach back to antiquity and the recognition of evident family resemblances among the various animal and plant species. In more recent history, Charles Darwin's grandfather, Erasmus Darwin, wrote a treatise, *Zoonomia; or, The Laws of Organic Life*, arguing for this idea in the 1790s. This view was widely debated and controversial. William Paley, in a highly influential book, *Natural Theology* (1802), argued that life, as complex as it is, could not have come into being without the help of a Designer. Numerous lines of evidence have become available in the two centuries since, through genetics and the fossil record, that persuade professional biologists that existing life forms on Earth are indeed related and have indeed evolved. This evidence contradicts Paley's conclusion, but it does not directly address his argument. A convincing direct counterargument to Paley's would need a specific evolution mechanism to be demonstrated capable of giving rise to the quantity and quality of the complexity now found in biology, within the time and resources believed to have been available.

The main contribution of Charles Darwin was, of course, exactly so motivated.[3] He posited the outlines of an evolution mechanism with two primary parts, namely variation and natural selection, that he argued was sufficient to explain biological evolution on Earth without a Designer. In its simplest form, the theory of natural selection asserts that each organism has some level of fitness in a given environment and that it is capable of producing a range of variants of itself as its progeny. It then attributes evolution to the phenomenon that among the variants, individuals that have characteristics that constitute greater fitness will have a higher probability of having descendents in later generations than those with less fitness.

Among biologists there is broad consensus that Darwin's theory is essentially correct. Biochemical descriptions of the basis of life provide a concrete language in terms of which the actual evolutionary path taken by life on Earth may one day be spelled out in detail and explained. At present there are many gaps in our knowledge. The relationship between the DNA (the genotype) and the behavior and physiology of the organism (phenotype) to which it belongs is little understood. In spite of this, over the last 150 years Darwin's theory has become the central tenet of biology by virtue of substantial other evidence. Most recently, DNA sequencing has given incontrovertible experimental confirmation for the proposition that the varied life forms found on Earth are genetically related. Nothing that I will say here is intended or should be interpreted as casting doubt on this proposition. However, it remains the case that Darwin presented only an outline of a mechanism. It is not specific enough to be subject to a quantitative analysis or to a computer simulation. No one has yet shown that any version of variation and selection can account quantitatively for what we see on Earth. There is much that needs to be explained. Evolution has found solutions to many difficult problems that are of value to life on Earth. These include, among many others, locomotion, vision, flight, magnetic navigation, and echo location. Humans have managed to find artificial solutions to these physical challenges only after enormous effort.

The achievements of evolution are palpable and objectively impressive. The possibility remains that every version of variation and selection, as we currently understand these terms, would have needed a million times as long to yield existing life forms than is believed to have been available. Saying that evolution is a contest or even a struggle for life does not go far in explaining these facts. No theory is known that would explain how competition by itself leads to such spectacular achievements. Lotteries, singing competitions, and gladiatorial contests have not produced similar improvements or novelty. Evolution is a special kind of contest. How are we to go about understanding how this special contest, of whatever kind it is, has been able to produce the spectacular inventions that it has?

The term evolution evokes many images—indeed almost all facets of the history of life on Earth. I will restrict attention here to the one primary question of how complex mechanisms can arise at all within the limited time scale and resources in which they apparently have. The numerous other questions that are widely discussed by evolutionary theorists I regard as secondary to this one. The advantages offered by sex to evolution have been much debated, but evolution was far along when sex arrived on the scene. The intellectual challenge of understanding how peacocks could have acquired their elaborate plumage was much troubling to Darwin. But again, peacocks came along late in the game. In short, what I seek to address is a gap between the general formulation of natural selection as currently understood and any demonstration that any specific mechanism can account for the biological evidence we see around us. Every scientific theory has a gap that leaves some question unexplained. Evolution is by no means unique in that respect. Having a gap is no fatal flaw. However, the natural selection hypothesis as currently formulated has the gaping gap that it can make no quantitative predictions as far as the number of generations needed for the evolution of a behavior of a certain complexity. I believe that the time is ripe for working toward filling this gap. And I believe computer science is the tool for doing it.

This may be an unconventional claim, but I will argue that Darwin's theory lies at the very heart of computer science. Darwin's theory may even be viewed as the paradigmatic ecorithmic idea. One of computation's most fundamental characteristics is the separation between the physical realization of a mechanism and its manifest behavior. This is equally true of Darwin's theory. Although the fitness of a biological organism depends both on the biochemistry of the organism and on all the physical, chemical, and ecological factors present in its environment, the principle of natural selection makes no mention of biochemistry, physics, or ecology, and it incorporates no specific knowledge about the fitness of a particular species in a particular environment. We are driven to the almost paradoxical conclusion that organisms that perform at such a sophisticated level of expertise in physics, biochemistry, and ecology are the products of generic mechanisms that incorporate no such expertise. This striking contrast summarizes the basic challenge that ecorithms in general, and evolutionary algorithms in particular, need to overcome.

Given the central role that Darwin's theory now plays in biology, the following fact is more than a little disconcerting. From the first availability of digital computers many intelligent, curiosity-driven individuals have sought to simulate selection-based evolutionary algorithms in order to demonstrate their efficacy. These simulation experiments, carried out over more than half a century, have been disappointing, at least in my view, in creating mechanisms remotely reminiscent of those found in the living cell. In fact, these experiments are seldom quoted as corroborating evidence for evolution.

This failure cannot be ignored. It suggests that the natural selection hypothesis has to be refined somehow if it is to offer a more explanatory scientific theory. Further, the refinement will need to have a quantitative component that reflects the realities of the actual bounded numbers of generations, and bounded numbers of individuals per generation, that apparently have been sufficient to support evolution in this universe. That evolution could work in principle in some *infinite* limit is obvious and needs little discussion. But modern humans are believed to have existed for no more than about 10,000 generations and with modest population sizes for much of that history. Our predecessor species may have had not dissimilar statistics. Theories of evolution that assume unbounded resources for evolution, in generations or population sizes, or those that do not address this issue at all, cannot resolve the central scientific question of whether some instance of natural selection does fit the constraints that have ruled in this universe.

I am not the first to point out that there is a tension between the long time apparently needed for evolution and the limited resources that evidence from the physical sciences suggests have been available. No one was more aware of this tension than Darwin himself. In an attempt to find corroborating evidence for the long time scale he believed was needed for evolution, he looked to geology. In the first edition of *On the Origin of Species* he included an estimate of 300 million years for the time needed for erosion to have created the Weald formation in southern England.[4] This estimate immediately came under fire from the scientific community. Darwin omitted it, and any other such estimate, from subsequent editions. William Thomson (later Lord Kelvin) and other authoritative physicists of his day derided Darwin's estimate as impossibly too high even for the age of the Earth itself. Their arguments were based on applying the principles of physics as then understood to the question of the rate at which the Earth had been losing heat. This indirect line of attack on his theory of evolution gave Darwin much reason for concern. He wrote, "Thomson's views on the recent age of the world have been for some time one of my sorest troubles."[5] Kelvin's final published estimate was as low as 24 million years.[6] Physicists now estimate the age of the Earth, thankfully, to be much higher, about 4.5 billion years (and 13.8 billion years for that of this universe). Nevertheless, we still do not have a quantitative explanation of how life could have reached its current state even within this more extended period that is now allotted by the physicists, whether on the Earth or in the broader universe.

The theory offered here, of treating Darwinian evolution as a computational learning mechanism and quantitatively analyzing its behavior, is the only approach I know that addresses these questions explicitly. Previous mathematical approaches to evolution, such as those of population genetics, analyze the effects of competition on relative population sizes. For example, the famous Hardy-Weinberg principle from the early twentieth century shows that if reproduction is sexual and members of a population have two copies of each gene, as in humans, then diversity in the gene pool will be conserved in the following sense. If two variants of a gene exist in the population in a certain ratio and they are equally beneficial, then their ratio of occurrence in the population will converge to a stable value, with both variants continuing to occur. Analyses of relative population sizes such as this, however, do not address how more complex forms can come into being from simpler ones—this is the most fundamental question and the one that opponents of evolution usually target. One is not performing a service to science if one pretends to have a solution when one does not.

Advances in biology over the last half century have made concrete what needs to be explained in ways that were not known to the earlier pioneers of population genetics such as the eminent statistician Ronald Fisher. We now know that biological organisms are governed by protein expression networks. To understand evolution we need to have an explanation of how such complex circuits can evolve from simpler ones and maintain themselves in changing environments. The protein expression networks on which our biology depends are known to have more than 20,000 genes, and the outputs they produce depend in a highly complicated way on the innumerably many possible input combinations. These circuits define how the concentration levels of the many proteins in our cells are controlled in terms of each other. We can seek to describe them mathematically. For example, the amount produced of our seventh protein may depend on the concentrations of three others—say, the third, twenty-first, and seventy-third. The dependence is something specific, perhaps $f_7 = 1.7x_3 + 3.4x_{21} + 0.5x_{73}$, or more likely something else. But in any case it is some particular dependency $f_7(x_1, \ldots, x_{20,000})$ on all the available proteins and possibly on some additional parameters, such as temperature. Whatever this dependency f_7 is, it will change during evolution if some other such dependency becomes more beneficial to the organism because of changing circumstances.

What an evolutionary theory must do is explain how these dependencies are updated during evolution. How long will it take to evolve to a new function f'_7 if the environment changes so that the new function f'_7 is better than the old f_7? Of course, this only accounts for evolution with a fixed set of proteins. A successful theory must also explain the evolution of new proteins. I believe that this will need a similar kind of analysis but for a different kind of circuit.

Over the last several decades it has emerged that there are computational laws that apply to the existence and efficiency of algorithms that are as striking as physical laws. These computational laws offer a powerful new viewpoint on our world that meets the challenge that the facts of biology lay down in regard to both evolution and learning. The laws that are most relevant to these phenomena are different from those that are the most useful for programmers of digital computers, and they need to be investigated separately. This will be our point of departure.

Nothing here is intended as the last word on any of the topics covered. The approach I propose needs extensive development both internally and in interaction with the experimental sciences it relates to. The idea that mathematical equations are useful for expressing the laws of physics, that laboratory experiments can uncover the facts of chemistry, and that statistical analyses in the social sciences yield clues about causation are all widely appreciated. But the notion that natural phenomena can be understood as computational processes or algorithms is much more recent. I have no doubt, however, that this algorithmic viewpoint is poised to take its place among the more familiar arsenal of weapons used for uncovering the secrets of nature. I hope to offer here a glimpse of how this algorithmic perspective will come to occupy a central position in science. First, however, we must turn to the questions of the nature and scope of computational processes in general.

Chapter Three

The Computable

Not everything that can be defined can be computed.

Computer science is no more about computers than astronomy is about telescopes.
EDSGER DIJKSTRA

3.1 The Turing Paradigm

In retrospect, humans have been remarkably uncurious for too long about information processing. Animals take complex inputs when seeing, smelling, touching, or hearing, and then produce behaviors that depend in complicated ways on these inputs. Human behavior can be even more perplexing and hard to understand. Phenomena like these we can observe every day. It would seem natural to wonder: Just how do living organisms process information and decide what to do? Curiously, until recent decades little intellectual effort has been put into understanding this question. To be fair to our predecessors, however, it is clear that, until recently, anyone attempting to study information processing would have been stymied by a fundamental impediment—no way was known of even formulating the question.

This only changed in the 1930s, when Alan Turing published a mathematical paper, "On Computable Numbers, with an Application to the *Entscheidungsproblem*," that inaugurated one of the most significant scientific revolutions in history.[1] The *Entscheidungsproblem* (or decision problem, in English) refers to a question raised by mathematician David Hilbert in 1928 concerned with deciding the validity of statements in mathematical logic. However, in his paper Turing went far beyond answering this one question. He formulated a notion that has changed how we view the world. Through its technological impact, this notion has changed how we live. His discovery was that computation, or the execution of step-by-step procedures for processing information, could be defined and studied systematically. Since that time we have been on a recognizable track toward understanding what such procedures can and cannot do. That is to say, we have come to understand computation. We have also been exploiting that understanding to produce technology, but technology is not my concern here.

The technical concept of computability makes an important distinction: It is one thing to specify, even unambiguously, what result you expect from a computation for every input of data. It is quite another to specify a step-by-step computation that gets you there. The difference is not immediately apparent. Nevertheless, Alan Turing proved that there exist problems for which there is no ambiguity as to what result is desired, but for which there is no set of step-by-step instructions that will get you the right result for every input. This was a stunning finding. Research over the past several decades has developed a rich science for making even finer distinctions, particularly with regard to efficiency. It turns out that there are also problems that are not computable efficiently enough to be practical, even if in principle they can be computed. That fact poses its own problems: We want computations not only to exist in principle, but also to deliver answers within a reasonable period of time. To obtain the result we should not have to wait for months, or years, or until after our galaxy has ceased to exist.

These laws of computation apply to all algorithms. Because ecorithms are algorithms, though of a special kind, they too must follow the same basic laws as computation in general. This new science of the ultimate limitations on the possibility and the efficiency with which computations for learning and evolution can proceed offers a fundamental new approach to understanding these phenomena of learning and evolution, because, regardless of how they are implemented—in silicon, DNA, neurons, or something else entirely—there are some ultimate logical laws that limit what these mechanisms can do.

Turing's paper contained several ingredients that are now seen as fundamental to the study of computation. First, he described a model, now called the Turing machine, that captures the phenomenon he was attempting to describe, namely that of mechanistic step-by-step procedures. Second, he proved a strong *possibility* result for what can be achieved on his model. In particular, he showed how to design a universal Turing machine that is capable of executing every possible mechanical procedure. This universality property is what enables computer technology to be so pervasively useful, and would be utterly astonishing were it not so commonplace now by virtue of its effectiveness. Third, Turing also proved a strong *impossibility* result, that not all well-defined mathematical problems can be solved mechanically.

Turing's impossibility result is as striking as universality is on the positive side. It is concerned with the problem of predicting, for an arbitrary computer program and an input for it, whether that program started on that input will ever halt its computation after a finite number of steps, as opposed to getting stuck in a loop in perpetuity. This so-called Halting Problem is well defined. Once we specify a language for expressing the programs there is no ambiguity at all about what would and what would not constitute a solution to it. It would be good to be able to tell ahead of time whether a computer program will get stuck in a perpetual loop. Yet, as Turing showed, it cannot be solved in all cases by any Turing machine.[2] We will never be able to solve this problem routinely.

Many of the foremost thinkers of the early part of the twentieth century had wondered, somewhat informally, whether mechanical procedures existed for resolving all mathematically well-posed questions. Some, such as the philosopher Bertrand Russell and the mathematician David Hilbert, were optimistic. Turing's discovery that one could define precisely what such an assertion meant, and then *prove* that such a statement was false, had revolutionary implications. The shock of this is still taking its time to permeate the community of the educated.

Important as the three particulars of Turing's paper are—namely Turing machines, universality, and noncomputability—they become even more significant when viewed as an instance of a general class of what I call a Turing triad: an unambiguous model of computation that captures some real-world phenomenon (mechanical calculation in Turing's specific case), and both possibility and impossibility results about that model. Learning, evolution, and intelligence are all manifestations of computational processes. As realized in nature, they may be subtle and operate near the limits of computational feasibility. We may need a correspondingly sophisticated understanding of computation before we can unravel their secrets. My

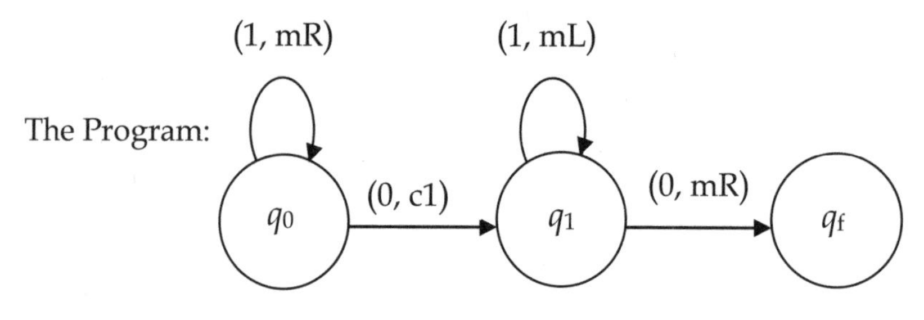

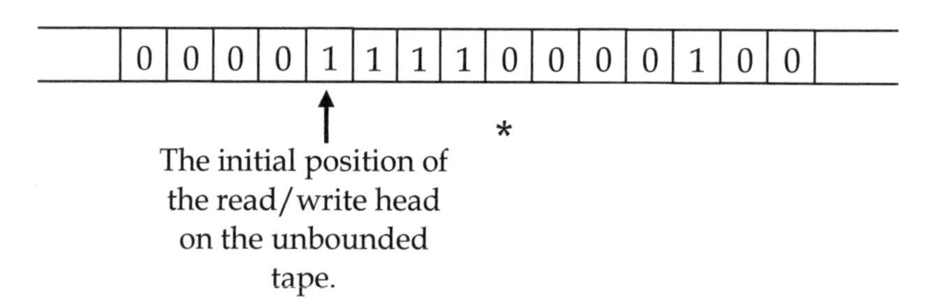

Figure 3.1 An example of a simple Turing machine. The diagram at the top describes the program that controls the machine. The input is the sequence of 0s and 1s on successive squares of the tape. The machine has three states q_0, q_1, and q_f. It starts in state q_0 and with the read/write head on the square pointed to by the thick arrow. If the machine is in state q_0 and the symbol under the head is 1 (as it is initially in this example), then the path indicated by the arrow out of the q_0 node with label starting with a 1 will be taken, in this case the arrow labeled (1, mR) with endpoint q_0. Executing this (1, mR) will result in the contents of the square being unchanged and the head moving one square to the right. The endpoint of the arrow indicates that the next state will be q_0 again. An arrow labeled (1, mL) would mean the same except that the head moves to the left. An arrow labeled (1, c0) would mean that the square is changed from 1 to 0 and the head does not move. The labels (0, mR), (0, mL), and (0, c1) have analogous meanings and apply when instead the symbol under the head is 0. The computation halts if and when a final state q_f is reached. The reader may verify by working through this example that, eventually, when the read/write head reaches the 0 at the * sign, the machine will change the 0 there to a 1 and change the state to q_1, then move the head back finally to the starting position, and then halt in state q_f. (Note that we can obtain an example of a machine that never halts on this input by changing the (0, mR) arrow from q_1 to go to q_0 rather than to q_f.)

strategy for shedding light on them will be to seek Turing triads for these phenomena also.

3.2 Robust Computational Models

The reader may have noticed that in the previous section there was an unexplained leap. The assertion that the Halting Problem was not computable by any Turing machine was identified with the claim that it was not computable by any conceivable mechanical procedure. To justify this leap, we will need a notion known as the robustness of models under variation, one of computer science's deepest and most fortunate mysteries.

We have seen that an essential ingredient of the Turing methodology is that of defining a model of computation that captures a real-world phenomenon, in this case that of mechanical processes, including those that no one had (or has yet) envisaged. That last part is crucial: With his machine, Turing aimed to capture all processes a human could exploit while performing a mental task that can be regarded as mechanical as opposed to requiring creativity or inspiration. The audaciousness of the attempt has attracted many who would prove Turing's machine insufficient to the power Turing claimed for it. However, when different individuals have tried to define their own notions of mechanical processes in hopes of creating models of greater power, all the models they have devised—no matter how different they may seem—could be proved to have no greater capabilities than those of Turing machines. For example, having two tapes, or five tapes, or a two-dimensional tape adds no new power. Similarly, allowing the program to make random decisions, or transitions that have the parallelism suggested by quantum mechanics, also adds no new capabilities. Extensive efforts at finding models that have greater power than Turing machines, but still correspond to what one would instinctively regard as mechanical processes, have all failed. Therefore there is now overwhelming historical evidence that Turing's notion of computability is highly robust to variation in definition. This has placed Turing computability among the most securely established theories known to science.

This robustness under variation of the model offers the fundamental key and launching pad for our study here. For learning and evolution, robust models are as indispensable as they are for general computation. Without this robustness the value of any model or theory is questionable. We are not interested in properties of arbitrary formalisms. We want some assurance that we have captured the characteristics of some real-world phenomenon. Robustness of models is the only known source of such assurance.

The discovery of the notion of computability constituted a new approach to discovering truths about the world. The logician Kurt Gödel generously acknowledged that computability theory "has for the first time succeeded in giving an absolute definition of an interesting epistemological notion, i.e., one not depending on the formalism chosen."[3] What can be computed does not change as one varies the details of the model. In later chapters, I shall try to persuade the reader that, for the same reason, analogous absolute definitions should be sought also for other notions, and in particular learning and evolution.

There is, of course, no reason to believe that for every notion for which there is a word in a dictionary there exists an absolute definition, or a robust computational model that captures its essence. Indeed, computability, learnability, and evolvability may be among the few. For most other notions no such robust computational models are known, and although robust models may be discovered one day for some, for the rest no such models may exist at all. The question of whether notions such as free will or consciousness can be made theoryful by the algorithmic method pursued here hangs, I believe, on whether robust computational models can be found for them.

3.3 The Character of Computational Laws

Turing's contributions amounted to more than a series of specific discoveries; they provided a new way of pursuing science. In this, his importance demands comparison with that of Isaac Newton. Newton's influence on physics is without parallel, not because he described gravity or made any other particular discovery, but because it was through his work that it became accepted that the physical world obeys laws that can be described by mathematical equations, and that solving these equations could yield accurate predictions of what will happen in the future. Newton's theories not only had the immediate generality that they applied very broadly to mechanical systems. They had a higher level supergenerality in that they offered a blueprint for developing theories for fields that had yet to be conceived. Physicists have followed this lodestone of expressing physical laws by mathematical equations ever since. Electromagnetic theory, general relativity, and quantum mechanics are not implied by Newton's mechanics, but they follow the same intellectual pattern: physical laws expressed as mathematical equations. In this sense, equations offered the wizardry that enabled successive generations of physicists to achieve an understanding of the physical world beyond that of which previous generations could have dreamed. Since the seventeenth century physics has been transformed several times as far as the range of phenomena that it could explain. Even as the particular discoveries of Newton have been superseded, physics is still being pursued with a methodology recognizably similar to that used by Newton.

No one knows why such supergenerality should exist in physics. It is sufficient for most purposes to recognize that it does. The physicist Eugene Wigner suggested that we simply enjoy its benefits: "The miracle of the appropriateness of the language of mathematics for the formulation of the laws of physics is a wonderful gift which we neither understand nor deserve. We should be grateful for it and hope that it will remain valid in future research and that it will extend, for better or for worse, to our pleasure, even though perhaps also to our bafflement, to wide branches of learning."[4]

Robust computational models, I expect, will turn out to provide supergenerality in computer science as mathematical equations have in physics. They will enable the extent and limits of computational phenomena, in all their variety, to be uncovered. Just as, in retrospect, the texture of all the discoveries in physics over the last three centuries can be recognized already in the work of Newton, the texture of much of the new science of the coming centuries will be traceable to Turing.

One can make some further observations regarding the two fields. Physics concentrates on understanding a minimal set of basic processes that are sufficient to explain the dynamics of the physical world, such as how particles move under natural forces. In contrast, computer science entertains much more diverse sets of processes—in fact, any process that can be formulated as step-by-step rules. As long as the trajectory of objects under the laws of physics can be simulated by step-by-step rules, as appears to be the case, computation will embrace all the processes studied in physics. However, computational processes, though more general than those of physics, are not totally arbitrary. They are governed by their own logical laws and limitations. The laws that govern them are our concern in the present chapter.

Prior to Turing, mathematics was dominated by the continuous mathematics used to describe physics, in which (classically, anyway) changes are thought of as taking place in arbitrarily small, infinitesimal increments. The Turing machine, however, is a discrete model. Before his time, discrete mathematics had been little explored or developed; in fact, a seldom discussed influence of Turing's work is the rise of discrete mathematics subsequent to it. It is striking that for the phenomena that we shall study here, including learning and evolution, discrete models again provide the most immediate robust models and have been most useful in isolating the basic phenomena. Continuous models are ultimately of at least as great interest, but for the initial explorations necessary to identify the most fundamental concepts they are not the most fruitful.

Besides the discrete versus continuous dichotomy, there is a more fundamental difference between physics and computer science. In physics we think of the equation as the immutable fundamental law, expressing such facts as that the gravitational force between two objects is proportional to the square of the inverse distance, and to no other function of the distance. In computation we have much broader latitude in constructing programs than this. We allow arbitrary programs composed of steps from some repertoire of basic steps. The immutable laws of computation are not constraints on how programs can be composed. Rather, like the noncomputability of the Halting Problem, they state what can or cannot be achieved by *any* program of a specified kind.

In computation the laws are statements, subject to mathematical proof or refutation, but their relevance relies on the robustness of the model in question. One may consider the laws of physics to be analogous to the laws of computation. However, as far as not being subject to mathematical verification,

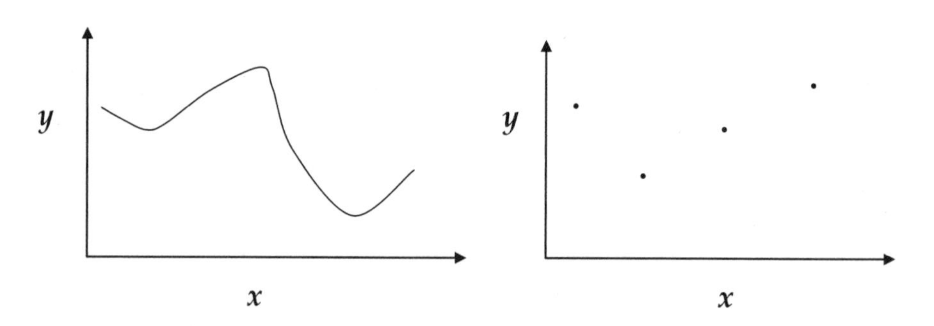

Figure 3.2 In a continuous model there are infinitely many possible states that are related to each other smoothly. The left-hand diagram shows an example where each state is represented by a point on the curve. In a discrete model the states need have no such relation. The right-hand diagram shows a model with four states, as indicated by the dots.

the laws of physics correspond in computation rather to assertions about the robustness of the models. The commonality between the laws of physics and robustness questions in computational models can be also stated positively—in both cases one needs to go to realities beyond mathematical formalisms for supporting evidence or falsification.

3.4 Polynomial Time Computation

Once computers had become more widely available and broader efforts were made to program them, the importance of understanding computational limitations in finer detail than computability theory provides came to the fore. The study of these limitations came to be known as computational complexity.[5] In that field one does not distinguish merely whether an algorithm for a specified task does or does not exist. One also quantifies how many steps any algorithm, if one exists, must take.

Using this idea, one can try to classify both familiar and unfamiliar tasks according to the number of basic operations that are required to perform them. The process of long multiplication for obtaining the product of two numbers is a familiar enough algorithm, taught in elementary schools worldwide. To find the product of two numbers, each of n digits in standard decimal notation, it takes about n^2 basic operations on pairs of single digits, as illustrated in Figure 3.3.

For long multiplication the actual number of operations on individual pairs of digits may be $4n^2$ or $5n^2$ or cn^2 for some fixed number c, depending on what exactly you consider an operation. It will not, however, grow faster than n^2, such as n^3 or n^4; nor will it grow more slowly, such as $n^{1.9}$. We can describe the order of growth using what is known as O notation, while omitting the less important detail of the value of c. We simply say that the long multiplication algorithm is an $O(n^2)$ algorithm.[6]

In general, we distinguish between an algorithm taking polynomial time versus one that takes exponential time. It is polynomial time if it takes $O(n^k)$ basic steps for some constant k, where n is the number of digits or bits needed to write down the input. Of course, it is best if k is a small number such as 1 or 2. An exponential time algorithm takes the form k^n (such as 2^n or 10^n). Exponential time algorithms become impractical even for moderate input sizes. For example, for a task taking 10^n steps, if n is just 30, then 1,000,000,000,000,000,000,000,000,000,000 steps are needed. A computer doing a trillion steps per second would take more than 30 billion years,

$$
\begin{array}{r}
314159265358979 \\
\times\, \underline{271828182845904} \\
1256637061435916 \\
0000000000000000 \\
2827433388230811 \\
1570796326794895 \\
1256637061435916 \\
2513274122871832 \\
628318530717958 \\
2513274122871832 \\
314159265358979 \\
2513274122871832 \\
628318530717958 \\
2513274122871832 \\
314159265358979 \\
2199114857512853 \\
\underline{628318530717958} \\
85397342226735418150399772016
\end{array}
$$

Figure 3.3 When performing long multiplication on two numbers each of n decimal digits, here $n = 15$, we multiply the first n-digit number by each of the n digits of the second number in turn, and then add the results. This can all be done by performing proportional to n^2 basic operations, additions and multiplications on pairs of single digits. However, these n^2 operations look repetitious. This raises the question of whether the same result can be achieved with far fewer operations. It is hard to explain why this very natural question had to wait till the 1960s to be asked and answered.

more than twice the currently estimated age of this universe, to accomplish this. Running many computers in parallel does not change the picture too much. If you had one computer for every particle in this universe, of which there are currently believed to be fewer than 10^{90}, then within this 30 billion years one could do $10^{90} \times 10^{30}$, or 10^{120}, operations. For a task taking 10^n steps, one could then solve instances of size $n = 120$. If we increased the speed of each computer by a factor of 1,000 we would increase the allowed input size only by 3, to get a limit on n of 123.

The point is that numbers such as 123 are very modest as input sizes. An input of 123 digits requires less than two lines on this page to write down. Most applications of computers need much larger inputs. For example, if one is scheduling a fleet of aircraft, then the input size will be in the thousands. If one is performing computations on data read in from the World Wide Web, perhaps analyzing articles written about some topic, then input sizes may be in the millions. In all such cases algorithms taking 10^n steps would be totally impractical. Even having all the resources of our universe at our disposal would be far from enough.

The class of tasks or problems that can be computed in polynomial time is represented by the capital letter P. The task of multiplying two integers is therefore a member of this class P, because the standard algorithm for it, as we have noted, takes $O(n^2)$ steps, which is polynomial. In general, P characterizes what can be computed in practice.[7]

Fundamental to computational complexity is the distinction between the outcome one wants to achieve, say, finding the product of two numbers, and the many possible ways of achieving it. Also fundamental is the notion that there exist easily specified problems that are computable in principle in the sense of Turing but for which all algorithms are impractically inefficient. The idea that we should classify tasks according to their computational difficulty appears to be very natural, and so this idea plays a central role in computer science. Yet there is an implication that goes a little against the grain of traditional science education. In conventional mathematics and science courses the computational tasks presented are invariably limited to those that are easy to compute, such as arithmetic and linear algebra. This tradition has the obvious justification that it presents only methods that are practical. However, a traditional education along these lines does leave the mistaken impression that *every* easily specified problem can be solved efficiently. It ill-prepares the student to face entirely novel challenges and find approaches to them that are computationally feasible. Turing's wartime work in breaking codes was centered on the problem of deducing from an encrypted message the original message, without performing an exhaustive search of the exponentially many possible keys that might have been used for encryption. Similarly, many natural tasks people would like to solve by computer, such as scheduling, involve finding the best solution from potentially exponentially many solutions. For many of these tasks exponential time algorithms are known, but none faster. Our scientific culture is still in the process of absorbing the significance of this phenomenon.

The impracticality of exponential time computations is self-evident. While the boundary between the practically computable and the infeasible is not sharp, the polynomial time criterion is the most convenient place that has been found to put that boundary. Clearly, polynomial time with high exponent, such as n^{100}, is as infeasible in practice as exponential time, even for modest values of n. However, the polynomial versus exponential distinction has proved very useful, simply because the majority of algorithms that are known for important problems conveniently dichotomize between feasibly low degree polynomials, such as quadratic, $O(n^2)$, and proper exponentials, such as 2^n. Hence, for reasons that are not understood, this polynomial versus exponential criterion is more useful in practice than its bare definition justifies. Experience shows that if someone claims to be able to compute a function routinely for arbitrary inputs of significant size, but claims that the problem is not in P, then there is a good chance that more can be and needs to be said. Perhaps the inputs are not really arbitrary but restricted to a special subclass or a probability distribution for which the problem is indeed solvable in polynomial time. More often than not, on further examination, one can explain such unexpectedly good performance. Indeed, much current research in computer science centers on the question of identifying the circumstances, sometimes one application at a time, in which polynomial time computation can be achieved in some useful sense, even if not in complete generality.

When defining computation, there is a further important distinction. A computation is deterministic if each step is uniquely determined from what has gone before. In the definition of P this is assumed. However, for all practical purposes we can relax this constraint of determinism to permit computations that make random choices as if they were tossing coins. These algorithms may still arrive at the correct answer with high probability, even if not with certainty. So-called randomized algorithms, ones that do arrive at correct answers with high probability for every input, are as effective as deterministic ones in practice, the probabilities involved arising only from the coin tosses the algorithm makes. Such algorithms give a wrong answer only for combinations of coin tosses that occur very rarely, such as getting heads only three times in a thousand tosses. Further, for every input, the probability that an error occurs can be driven down to be exponentially small by simply repeating the algorithm enough times, the probability of error being independent for each repetition. One can extend the definition of standard deterministic Turing machines to allow them to make decisions according to the toss of a coin in this way. These are called randomized Turing machines. The corresponding polynomial class is called BPP, for *bounded probabilistic polynomial* time. It is possible that P and BPP are mathematically identical, in which case every computation that uses randomization could be simulated in polynomial time by one without it. This question of whether P and BPP are equal is currently unresolved.

A class broader still than BPP is called BQP, for *bounded quantum polynomial* time. This class is inspired by quantum physics, which posits that a physical system can be in multiple states at the same time, in a certain specific sense. It is natural to ask whether such quantum phenomena can be exploited to speed up computation. Oversimplifying a little, quantum phenomena may permit a million computations to be pursued simultaneously in a single quantum computer, while conventional computers would need to

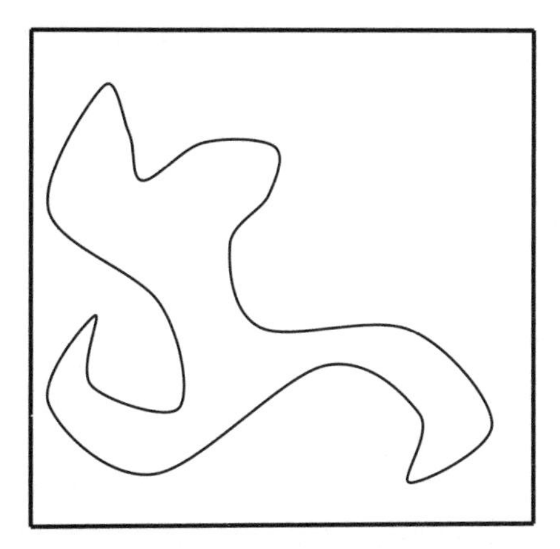

Figure 3.4 An example of a randomized algorithm. One can estimate the area of any shape by drawing it in a square of known area, throwing darts randomly at the square, and counting what fraction fall within the shape. This will work for any shape. All that is required is that the darts have uniform probability of hitting any part of the square, and that the successive throws be independent of each other. The only risk of getting a bad approximation of the area is that of being unlucky and getting throws that are not representative of uniformity. The probability of such an outcome goes down as one throws more and more darts. Randomized algorithms and the class BPP have essentially this guarantee of success.

do these one after the other in a million phases on a single machine, or in parallel on a million machines. Much effort has been expended to understand the power of polynomial time quantum computation. On the one hand, one would like to understand better, in mathematical terms, the power of the restricted parallelism that quantum theory seems to offer. On the other, one would like to know whether such machines can be constructed at all in this physical world, since quantum computers require certain capabilities that have not been shown to be realizable.

One can define the class PhysP to be the maximal class of problems that the physical universe we live in permits to be computed in polynomial time. Identifying the limits of the class PhysP would appear to be one of the great scientific questions of our time. BQP is a natural candidate. If it turns out not to be realizable, then BPP is the most natural alternative known.

While identifying PhysP is fundamentally a question about physics, mathematics may have a role in resolving it. It is possible that one can prove by a mathematical demonstration that P = BPP or BPP = BQP or even P = BPP = BQP. This last eventuality, for example, would show once and for all that polynomial time quantum machines have no more power than polynomial time versions of the standard deterministic machines defined by Turing that was illustrated in Figure 3.1.

One thing we do know is that each of these three classes—P, BPP, and BQP—itself has substantial robustness under variation. Attempts to characterize deterministic, randomized, or quantum computations have yielded just one good candidate computational model for each class. This robustness for the randomized and quantum classes is known only for problems with yes/no answers. This currently leaves two main candidates, BPP and BQP, for the practically computable yes/no problems in this physical world. Having two candidates is only a small embarrassment, further alleviated by the fact that the range of natural problems that have been identified to be in BQP but are not known to be in BPP is somewhat limited.

If we leave aside the constraint of polynomial time, of course, any of the algorithm types—deterministic, randomized, or quantum—is provably as good as any other. The characteristic robustness of Turing machines remains. In the opposite direction, if we instead impose more and more constraints, beyond the polynomial constraint, to reach the learning and evolvability classes that we shall meet later, the robustness criteria become increasingly challenging to satisfy.

3.5 Possible Ultimate Limitations

The Turing methodology that we described earlier when applied to a specific task such as integer multiplication would consist of the following three components. Define an appropriate model that captures the realistic cost of computing the task. Prove possibility results, in this case efficient algorithms for the task that take few steps. Prove some impossibility result that shows, for example, that for the model defined no algorithm exists that takes fewer than so many steps.

There are many problems that we would wish to solve efficiently but do not know how. The most efficient algorithms known for a large class of these problems take an exponential, rather than a polynomial, number of steps. Certainly, there is no necessary reason why the best currently known algorithm should be the best possible algorithm. Let's return to the problem of multiplication. The basic question is this: What is the most efficient method for multiplying two n-digit numbers? This question has inspired a research program that has been pursued for half a century now. In 1960 an initial algorithm that took only $O(n^{1.6})$ steps was discovered by Anatoly Karatsuba, working in the Soviet Union.[8] For large values of n, this already improved substantially on the classical $O(n^2)$ method. It is more than a little surprising that this discovery, that integers could be multiplied much faster than by the standard method taught to children worldwide, came so recently.

After this initial discovery there rapidly followed a sequence of improvements. These culminated in the algorithm published by Arnold Schönhage and Volker Strassen in 1971.[9] This had runtime close to but not quite linear—that is, $O(n)$—but better than $O(n^{1+x})$ for any $x>0$. As a result of these developments we now know that integer multiplication is easier to compute than anyone would have had reason to suspect in earlier centuries, when only $O(n^2)$ methods were known and nothing better suspected.

To our embarrassment, we do not yet know whether multiplication is substantially more difficult than addition, which can be done by the standard method in linear time. Ironically, there has been little (in fact close to no) progress on establishing such an impossibility result. It is clear that any algorithm would need to look at all the $2n$ digits of the two input numbers, and hence that this computation cannot be done in fewer than $2n$ steps. However, the possibility remains that there exist linear time algorithms for multiplication, say $10n$ operations on pairs of one-digit numbers, as there are for addition. Resolving whether such a linear time algorithm exists for integer multiplication with inputs and outputs represented in the standard decimal or binary notation remains a major challenge for theoretical computer science. Is multiplication an inherently harder task than addition, or does it just appear to be so?

Multiplication is a comparatively simple problem, and clearly already computable in practice by means of the ancient algorithm. The question of whether polynomial time algorithms will be found someday for any of the many problems for which we currently have only exponential time algorithms is addressed in the field of complexity theory. We shall now review some of these results in the remainder of this section. The reader may find this interesting background in computer science, but it is not indispensable for what comes later.

A celebrated class of problems is the so-called NP, or nondeterministic polynomial time, class. These are characterized as the problems for which solutions may or may not be hard to find but for which a candidate solution is easily verified. For example, suppose we want to know whether a target number x, say 923, can be factored as the product of two smaller numbers p and q. Then for any given candidate pair p, q we can easily verify whether or not they are the factors of x simply by multiplying them together and checking whether the answer equals x. (For example, given the candidate numbers 71 and 13, it is easy to determine whether or not 71×13 is equal to the target 923.) But given just the number 923, there is no similarly easy route known to discovering the 71 or the 13. One naïve method for discovering such factors would be to enumerate all numbers less than the target x and test each one for whether it divides x exactly. This would, however, take about 10^n steps for n digit numbers. (The best method currently known for finding the factors of n-digit numbers is exponential in the cube root of n, which is a considerable improvement, but still not polynomial.)

The primality problem is the problem of determining for an arbitrary number x whether it has factors other than itself and 1. It is an NP problem since this verification of a particular candidate solution can be done in polynomial time (i.e., as we have observed, given an n-digit number x and two further numbers p and q, we can verify whether $pq=x$ in $O(n^2)$ steps). It turns out that, in fact, there do exist some very clever algorithms that can determine in polynomial time whether a number is prime. They reveal whether factors *exist*, but, curiously, not what these factors are.[10] Hence this particular NP problem of determining primality is in fact also in P.

Currently, there is no way known for finding the factors in polynomial time, even with randomization (BPP). The apparent exponential gap for classical computation between the difficulty of *testing* whether factors exist and *finding* them if they do is the basis of widely used cryptographic schemes, notably the RSA cryptosystem.[11] In the RSA system you choose two large prime numbers p and q, and multiply them together to get their product x. You then make public only the result x, and keep p and q secret. Anyone in the world who sees x can encipher messages intended for you, but only you, who know p and q, will be able to decrypt any such message. The point is that generating arbitrary primes p and q requires only the generation of some random numbers and testing whether they are prime. The eavesdropper needs to do the apparently much harder task of actually finding the factors of a particular x. (This factoring problem is known to be in BQP, or computable in polynomial time on a quantum Turing machine. This fact lends some intrigue, at least, to the question of whether it is feasible to construct quantum computers.)

The importance of NP is that it captures the very general process of mental search.[12] We call these problems mental search because they can be solved by searching objects one generates internally in one's head or computer. They do not require searching in the outside world, as one would when searching for a phrase in the World Wide Web or for oil in the ground. Given a particular problem, one can characterize a set of potential solutions large enough that any true solution must be in that set of candidates. For the problem of determining whether some n-digit number x can be factored, one may specify the potential solutions as the integers $\{2, 3, \ldots, x-1\}$. Finding the solution is simply a matter of testing each number, one by one, to see whether it divides x. Such exhaustive searches are not feasible for large values of n, for this or any other problem. For any NP problem the crucial question therefore is whether a more efficient process for detecting the existence of solutions than such an exhaustive search is possible.

The primality problem does have such a fast alternative algorithm, but it is by no means typical of NP problems. For a very large class, and one could say for the majority of natural NP problems, no algorithm is known that puts them in P or BPP or even, like the factoring problem, BQP. Remarkably, it has been shown that all the members of a very large class of NP problems are in fact provably equivalent to each other, in the sense that a polynomial time algorithm for one would give a polynomial time algorithm for any other. This class has the further remarkable property that each member is provably the hardest member of NP. In other words, for these so-called NP-complete problems, no one currently knows a polynomial time algorithm for any of them, but if someone did find such an algorithm for *any* one, then polynomial time algorithms would follow for *all* problems in NP.[13]

An example of such an NP-complete problem is the Traveling Salesman Problem. Here one is given a map containing some cities, the distances between the pairs of cities that have direct roads between them, and a number x. The problem is to determine whether there is a tour that traverses every city exactly once and has total distance no more than x. This problem is in NP because given a candidate tour it is easy enough to verify that each city is traversed exactly once and that the total length of roads used in the tour is less than x. Scheduling problems in all their variety offer a host of other mental search tasks for which we wish we had efficient algorithms, but we now know that most are NP-complete. NP-complete problems can be found in every area of mathematics. For example, for algebraic equations we have the primality question as to whether, given an n digit integer c, the equation $xy=c$ has a solution in integers x and y. Of course some equations are easier to solve than this. Given integers a, b, and c, whether the equation $ax^2+bx+c=0$ has an integer solution can be solved by the standard formula for solving quadratic equations. On the other hand, the superficially similar question of whether the equation $ax^2+by+c=0$ has any integer solutions x, y, is NP-complete![14] Yes, we are told only about the easy things in high school.

Because intensive efforts to find polynomial time algorithms for NP-complete problems have to date failed, many currently conjecture that $P \neq NP$, or equivalently, that no polynomial time algorithm exists for NP-complete problems. (NP-complete problems are similarly conjectured not to be in BPP or QBP either.) Whether this conjecture is in fact a computational law, like Turing's proven assertion that the Halting Problem is not computable, is potentially resolvable, and I expect that it will be proved or disproved one day. The postulate $P \neq NP$, while it remains unresolved in either direction, might be compared to laws in physics, which likewise have not been mathematically proven. Of course, a physical law cannot be proven by mathematics. Such computation postulates can play analogous roles to physical laws in the sense that we can make good use of them as working hypotheses, at least until someone disproves them. In this instance the working hypothesis is that polynomial time algorithms do *not* exist for NP-complete problems. The eventuality that this is disproved could, of course, be a very happy one if it is accompanied by the discovery of an efficient algorithm for all NP-complete problems, which would have revolutionary consequences if it was efficient enough.

In later chapters as I move on to consider learning, reasoning, and evolution, I shall seek to follow the Turing triad: establishing a robust computational model, proving some strong possibility results, and proving some impossibility results that explain the ultimate limitations. As in complexity theory generally, proving impossibility results is particularly challenging. We may need to be ready to postulate certain algorithmic laws, in analogy with NP-completeness, without being able to prove them. Such postulates can then be treated as working hypotheses, at least until someone disproves them and finds, unexpectedly and pleasantly, that a whole range of computational phenomena, currently believed to be infeasible, is indeed feasible.

A potentially wider class of computations still than NP is #P (pronounced "sharp P"). This is the class of problems that enumerate the number of solutions of NP problems. They give a number as the output. This class also has a class of its hardest members, called the #P-complete problems, analogous to the NP-complete problems. It is clear that for an NP-complete problem counting the number of solutions is at least as hard as detecting whether there are any, since the answer will be a number, and if it is greater than zero, then we will know that there exist solutions. More interestingly, there are many natural problems where testing whether there exists a solution is in P but counting their number is #P-complete. This means that while the existence of solutions can be detected fast, counting the number of solutions is as hard as for NP-complete problems.[15] Examples of such problems abound in the context of reliability—for example, where one wants to determine the probability that a complex network or system will fail from the failure probabilities of the components. Since the probability of something happening is closely related to the number of ways it can happen, these problems can be viewed as counting problems. It turns out that this class #P is at least as powerful as not only NP but also the quantum class BQP.[16] It remains a possibility therefore that a yet undiscovered polynomial time algorithm exists that computes all problems in #P, and hence also all problems in BPP, BQP, and NP.

The importance of these complexity classes derives from the additional fact that they are useful for classifying naturally occurring problems. Many problems that arise are mental search problems or their corresponding counting problems. It just so happens that when we come across a new task that we would like to have solved, if we cannot find a polynomial time algorithm for it, then more often than not, we can prove that it is complete in (i.e., is a hardest member of) its class NP or #P. Logically they could fall in

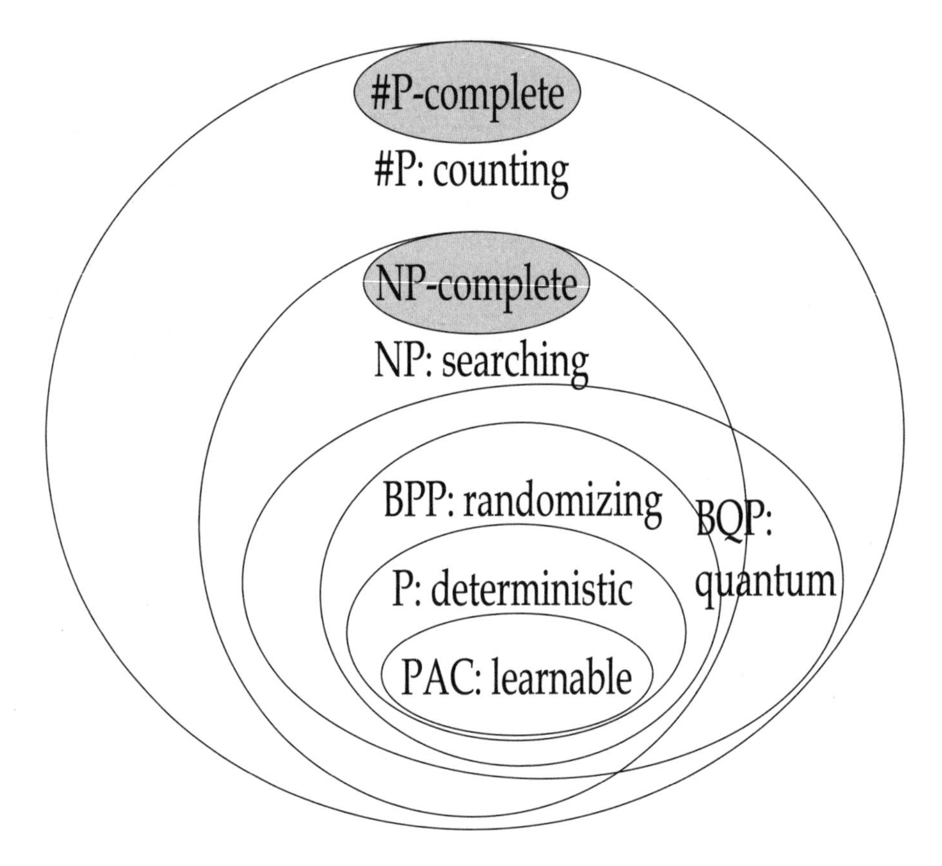

Figure 3.5 An illustration of the relative computational power of some complexity classes, as understood in 2013. Each ellipse represents a class of problems or tasks. Each point in each ellipse represents a problem, such as testing numbers for being prime, or the Traveling Salesman Problem.[17] The first glimpse that natural problems were related in this elegant way was given in a historic paper published by Stephen Cook in 1971 that defined the NP-complete class. The diagram illustrates the previously unsuspected rich structure that is now known to abound among different problems. The PAC class represents the feasibly learnable and is the subject of Chapter 5.

between, but for reasons we do not understand, they rarely do. For this reason this theory gives useful guidance as to the practical solvability of new problems as they arise. Why natural problems should dichotomize in this way is not predicted or explained by any known theory. It is one of those Wignerian mysteries that we neither understand nor deserve, but should be grateful for and simply enjoy.

These basic questions, about the relative extents of these complexity classes, are related intimately to the question of the real extent of PhysP, the class that this universe *physically* permits to be computed efficiently. If the quantum class BQP turns out to equal PhysP, we still would like to know whether NP or #P is within that class, and hence also permitted by physics. Questions about the relative power of these complexity classes can be viewed therefore as questions of physics also.

3.6 Simple Algorithms with Complicated Behavior

Ultimately, as we have seen, there are some limitations on what algorithms can do. Another way of saying this is that our powers to specify what we wish to compute are greater than the expressive power of computation itself. Nevertheless, the language of algorithms, despite these limitations, can be itself very expressive. Turing's result that there exist universal Turing machines that can simulate any computation is a clear statement of the breathtaking power of algorithms, the algorithm in question there being the one that controls the universal machine.

That, of course, we have already seen. A different facet of the richness of algorithms is that even some very simple specific cases can have behaviors that are mystifying to mere mortals. A well-known example of a simple procedure that has so far defied analysis is the following:

1) Start with any positive integer n.
2) Repeat until $n = 1$:
   (a) If n is even, replace n with $n/2$.
   (b) If n is odd, replace n with $3n + 1$.

For example, starting from $n = 44$, we get the following sequence: 44, 22, 11, 34, 17, 52, 26, 13, 40, 20, 10, 5, 16, 8, 4, 2, 1. For a fixed starting point, such as $n = 44$, computing the successive members of the resulting sequence is easy enough. What is not known is whether the sequence generated for *every* starting point n eventually reaches the value $n=1$ and terminates. Many starting points have been tried since the mathematician Lothar Collatz posed the problem in 1937. They all resulted in computations that did terminate at $n=1$. But—somewhat shockingly, given how simple the problem is to describe—no one has been able to offer a proof that this process would terminate for every possible starting point, or that it would not.

Collatz's problem is an example of apparent inherent complexity in simple procedures, even those isolated from any complex environment. In this case the notion of input can be removed altogether by considering a compound procedure that feeds the starting numbers $n=2, 3, 4$ in succession to the basic procedure, going on to the next starting number when the sequence generated by the previous one has terminated at $n=1$. Asking whether this compound procedure will ever get to every starting number n, rather than get stuck in perpetuity after a specific n, is equivalent to the original problem. In this light we should not be so shocked by the noncomputability of the Halting Problem, which would need to be able to make some kind of prediction about the ultimate fate not just of one, but of any computation.

3.7 The Perceptron Algorithm

Our journey through the major themes of computational complexity now brings us finally to the vicinity of our destination, the study of ecorithms. Our final point of departure is a simple but important algorithm that, like Collatz's problem, can also have complicated behavior, but these complications can be attributed to the outside environment in which it operates. This example is the perceptron algorithm, proposed by Frank Rosenblatt in the 1950s.[18]

The perceptron algorithm operates in the following context. Assume that there is a set of potential examples, each one specified by some description, and further that there is a criterion for which some of the examples are true examples and the others false. For instance, an example may be an individual flower, and the criterion may be whether that flower belongs to species A or species B.[19] The perceptron algorithm requires that the examples be described somehow. For this case, let us say that the description consists of two numbers x and y that specify the length and width of one of its petals.

The perceptron algorithm is a member of the class of supervised learning algorithms, which means it can be trained to do the work of classifying examples according to our criterion and descriptions. First, the algorithm is given descriptions of a set of training examples, as well as the correct label of each. For instance, one flower may have a petal 3 units long and 1 unit wide, and be labeled as a member of species *A*. In a subsequent phase the algorithm is fed with a set of test examples, which consist of descriptions of examples but no labels. The goal of the algorithm is to predict reliably for each test example whether it is true or false, or as in the flower case, an instance of species *A* or species *B*.

The perceptron algorithm works when there is a certain mathematical criterion, known as a linear separator, dividing the two possible classes. This criterion, in the case of our flowers, is a rule of the form

$$px + qy > r$$

where p, q, and r are numbers, such that every flower that satisfies it is of species *A*, and every one that does not is of species *B*. For example, suppose that $p = 2$, $q = -3$ and $r = 2$, so that the rule is

$$2x - 3y > 2.$$

Then a flower with petal length 5 and width 2 would be classified as type *A* since $(2 \times 5) - (3 \times 2) = 4$, and $4 > 2$. On the other hand a flower with petal length 3 and width 2 would be classified as type *B* since $(2 \times 3) - (3 \times 2) = 0$ and $0 < 2$. In graphical terms this means that if all the examples are plotted in two dimensions, representing the length by x and the width by y, then there is a straight line corresponding to equation $2x - 3y = 2$, so that all the species *A* flowers lie on one side of this line, and the species *B* flowers on the other (or on it). This is illustrated in Figure 3.6.

Of course, the perceptron algorithm does not know the true equation for the separator in advance. Instead, it must find it. The algorithm works by scanning through the training data, possibly many times. At each instant it maintains a hypothesis, of the form of $ax + by > c$, about the linear separator. We shall for simplicity work with the case $c = 0$.[20] The algorithm then starts with the hypothesis $0x + 0y > 0$. It goes through each training example one by one, and if the example label is correctly predicted by the current hypothesis, then the hypothesis is not changed. If the example label is not predicted correctly, then the hypothesis is updated so as to be "more likely,"

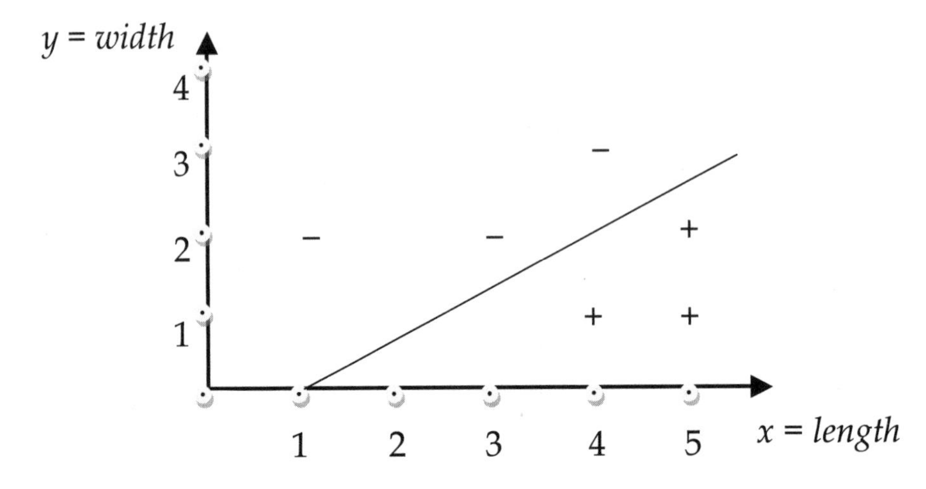

Figure 3.6 The sloping line contains the points satisfying $2x-3y=2$. The points $(x=4, y=1)$, $(x=5, y=1)$, and $(x=5, y=2)$ all satisfy $2x-3y>2$ and are marked as "+," while the points $(x=1, y=2)$, $(x=3, y=2)$, and $(x=4, y=3)$ do not and are marked as "–." In other words, the flowers of species *A* will lie below the line, and those of species *B* above or on it.

in a certain sense, to be correct on that same example if presented again later.

To be more precise, if the hypothesis misclassifies a true positive example (u, v) as negative (i.e., because $au+bv\leq 0$), then a is updated by having u added to it, and b by having v added to it. The left-hand side of the updated hypothesis will then be $(a+u)x+(b+v)y$, and it will have value $(a+u)u+(b+v)v$ if presented with the same example (u, v) on a subsequent run through the data. The value of the sum will be larger than before by a positive quantity u^2+v^2, and hence will be "more likely" to exceed 0 in value and correctly identify the positive example as true. For the opposite case, when a negative example is misclassified to be positive, a is updated by having u subtracted from it, and b by having v subtracted from it. This will have the effect of reducing the value of the left-hand side by u^2+v^2 if the same input (u, v) is presented again later, and hence will make it "more likely" to be less than 0 and hence result in a correct negative classification in that eventuality.

Depending on the order in which training data is fed to the perceptron, it could generate exceedingly many distinct histories of hypotheses. The interesting fact about the perceptron algorithm is that, in spite of our lack of control over its exact fate as we let it loose on arbitrary data, it nonetheless manages to achieve something quite remarkable. The most basic statement of the power of this algorithm, proved by Albert Novikoff soon after the algorithm was first proposed, is that if there is a true linear separator, then the algorithm is sure to find it, or another hypothesis that also correctly classifies all the examples, after having made misclassifications only a finite number of times. Furthermore, an upper bound on the number of such misclassifications can be computed given the data. This upper bound is equal to M/m^2, where M is the square of the distance of the furthest data point in the training set from the point (0, 0) and m is the margin. The margin has a trickier definition: It is the minimum distance of any data point from the separating line for the line for which this distance is the largest. The consequence of m being in the denominator is that the closer the data points are to the separator, the more mistakes this procedure can potentially make.

The importance of the algorithm derives from several additional facts. First, it works within bounds of the form M/m^2 not just for problems with two variables but for problems with any number of variables. Second, in practice, it often works well even for data that is corrupted by noise. Third, there are general methods for dealing with data that is separable not by linear relations but by more complex curves. For example, suppose that the two categories are not separated by a straight line as they are in Figure 3.6, but we suspect that some more complex curve would separate them. In our two-dimensional case we could try to learn the separator $ax + by + cxy + dx^2 + ey^2 > f$ where x, y are variables and a, b, c, d, and e are the constants to be learned. This inequality is not linear in x, y, since it contains higher order terms such as x^2. However, it can be viewed as linear if we regard the set of variables not as $\{x, y\}$ but as $\{x, y, xy, x^2, y^2\}$. We can translate any example given as a pair $\{x, y\}$ of numbers to the corresponding five numbers $\{x, y, xy, x^2, y^2\}$ by multiplication. In this way the perceptron algorithm can be applied directly to nonlinearly separable data also.

This linearization is an important idea that greatly extends the range of applicability of the perceptron algorithm, but it is not the complete panacea that it may seem. If there are few nonlinear terms, and we know which they are, then there are no problems. But if there are numerous terms potentially to look for, then this will introduce higher, possibly exponential, costs.

The criterion that only a finite number of mistakes are made over any, even infinite, number of examples does not appear to be a natural fit for human learning. It raises the question of what outcome we should really require

True Value	Example	Classification by Previous Hypothesis	Updated Hypothesis
			$0x+0y+0z>0$
+	(4, 1, 1)	–	$4x+1y+1z>0$
–	(1, 2, 1)	+	$3x-1y+0z>0$
+	(5, 1, 1)	+	$3x-1y+0z>0$
–	(3, 2, 1)	+	$0x-3y-1z>0$
+	(5, 2, 1)	–	$5x-1y+0z>0$
–	(4, 3, 1)	+	$1x-4y-1z>0$
+	(4, 1, 1)	–	$5x-3y+0z>0$
–	(1, 2, 1)	–	$5x-3y+0z>0$
+	(5, 1, 1)	+	$5x-3y+0z>0$
–	(3, 2, 1)	+	$2x-5y-1z>0$
+	(5, 2, 1)	–	$7x-3y+0z>0$
–	(4, 3, 1)	+	$3x-6y-1z>0$
+	(4, 1, 1)	+	$3x-6y-1z>0$
–	(1, 2, 1)	–	$3x-6y-1z>0$
+	(5, 1, 1)	+	$3x-6y-1z>0$
–	(3, 2, 1)	–	$3x-6y-1z>0$
+	(5, 2, 1)	+	$3x-6y-1z>0$
–	(4, 3, 1)	–	$3x-6y-1z>0$

Figure 3.7 Example of a run of the perceptron algorithm in three dimensions on the set of six examples +(4, 1, 1), –(1, 2, 1), +(5, 1, 1), –(3, 2, 1), +(5, 2, 1), –(4, 3, 1) repeated in that order three times. The signs indicate the labels of the examples. The initial hypothesis is $0x+0y+0z>0$. The first example (4, 1, 1) when substituted in the left-hand side of the initial hypothesis gives 0, and hence does not satisfy it, as indicated by the negative sign in the third column. The first column indicates that the true label of this first example (4, 1, 1) is positive. The algorithm therefore adds the coordinates (4, 1, 1) of the example to the coefficients (0, 0, 0) of the hypothesis, to give $4x+1y+1z>0$ as the updated hypothesis. After the six examples are cycled through twice, the hypothesis $3x-6y-1z>0$ is obtained. In the third cycle it is confirmed that this hypothesis satisfies all six examples.

of a learning algorithm before we declare it successful. This is the main question that will be addressed in Chapter 5. Before we get there, however, we need to take a more general look at what a computationally sound, mechanistic explanation of a natural phenomenon—whether of evolution, or cognition, or some other process of interest—might look like.

But there is one intuition suggested by the perceptron algorithm that will be important for what comes later. Learning is achieved in many steps that are plausible but innocuous when viewed one by one in isolation. These steps work because there is an overall algorithmic plan. In combination the steps achieve something, in particular, some kind of convergence. We shall claim that evolution is similar. The many small steps taken do not make too much sense one by one. But there is an algorithmic plan, so that taken in unison the many steps do achieve something remarkable.

Chapter Four

Mechanistic Explanations of Nature

What might we look for?

> *I hope it will not shock experimental physicists too much if I say that we do not accept their observations unless they are confirmed by theory.*
> ARTHUR EDDINGTON[1]

On February 28, 1953, Francis Crick announced to the patrons of the Eagle pub in Cambridge, England, that he and James Watson had discovered the "secret of life." What they had discovered was the double-stranded helical structure of DNA, the molecule that by then was suspected to be the carrier of heredity. This structure, with the two strands containing identical information, was suggestive of the process by which cells might copy their DNA during replication. The two strands simply separate, each strand carrying all the information it needs to give rise to a new double-stranded version of itself in its own new cell.

Something that became alarmingly clear after the content of the human genome had become largely known is that knowledge of the DNA sequence does not by itself unlock all the secrets of life. The sequence specifies the circuits of human biochemistry, but in a code we understand only partially. More than half a century after Crick and Watson's discovery, we still know little about how knowledge of the sequence can be exploited to understand the physical processes inside the living cell or to help cure disease. Despite everything that we know we do not know, we do have some insight into the computational nature of DNA. A strand of DNA consists of a sequence of nucleobases, each of which is one of four different chemicals, adenine, guanine, thymine, and cytosine. The sequence of bases contains the information that is carried by the living cell and inherited by its offspring. It may seem elementary, but it is still noteworthy that the way the information is represented in DNA is the same as it is represented in a Turing machine, as a sequence of symbols from a fixed alphabet. In the case of DNA the alphabet has the four symbols A, G, T, C, standing for the four nucleobases. And as Turing showed, a one-dimensional sequence of symbols from a fixed finite alphabet can describe and support all computations.

That the information in DNA is stored in a sequence is simply the first and most immediate of the many ways in which biology may be viewed as computational. When the cell divides, the base sequences are scanned for copying much like Turing machine tapes are during computations. Random mutations are realized by bases changing one to another just like randomized Turing machines would change a symbol. Since errors may be made in copying, methods are also needed for correcting errors.

The operations carried out in living cells and larger structures, such as our neural networks, can be usefully viewed as computations at many deeper levels as well. One is at the level of the protein expression circuits that the DNA sequences define. At any one time some of the proteins are expressed (produced) in the cell, and these in turn cause other proteins to be expressed according to the interdependencies specified in the protein expression circuit. On a different scale, the nervous system can be viewed equally as a very large circuit that performs elaborate computations that we as yet also understand only a little.

We can also ask the higher level question of how these protein or neural circuits are themselves created and maintained. Evolution is realized by modifications in the DNA sequences and hence in the protein circuits. These modifications can be regarded as computations also. With regard to neural networks, organisms learn during life by adapting their neurons in response to events. These adaptations are again computations.

An early example of a computational view of biology was given by Turing himself, in his theory of morphogenesis, or the development of shape. This has had considerable influence on thinking about how the many cells of the embryo can differentiate themselves and take up their various roles in a complex organism, despite having arisen from one unspecialized cell. Among other things, Turing suggested that the wide variation of the dappled patterns on animal fur, whether Dalmatians or leopards, might be accounted

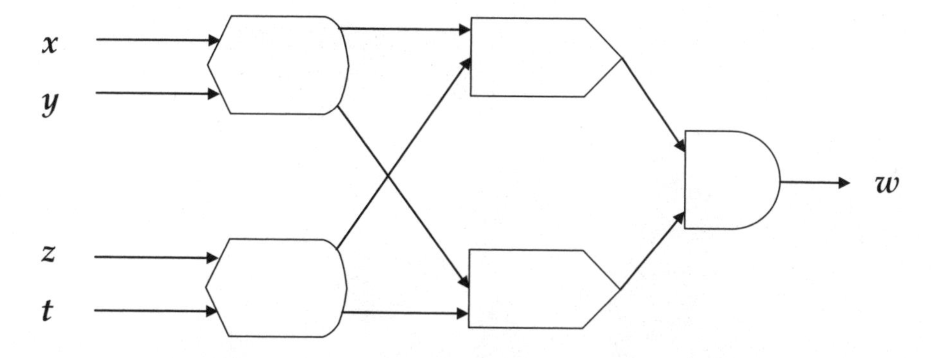

Figure 4.1 An illustration of a circuit. A situation is described by values *x*, *y*, *z*, and *t* that are input to the circuit. The value of the response of the circuit is *w*. Each circuit component performs some operation on the input values or on the results of previous operations. A circuit can be regarded as a general computation where the dependence among the various inputs, outputs, and intermediate computed values can be made explicit, as in this diagram. A neural or protein circuit will be effective if its response is beneficial to the owner of the circuit in typically encountered situations. For theoryless decisions it is sufficient that the circuit be effective in situations that are most frequently encountered by the owner—no theory or understanding of why it is effective is needed. Ecorithms are the mechanisms by which such circuits are acquired and kept in tune.

for by random variation during development, even if the animals are genetically all identical. Turing demonstrated his suggestions by simulations (as shown in Figure 4.2). He was giving one of the earliest examples of computational science, the idea that facts about the world can be discovered not just by physical experimentation or by positing theories, but also by computational simulations. Such simulations can sometimes pursue the consequences of hypothesized theories beyond where mathematical analysis is able to go.

Biology therefore is based on complex mechanisms at many different levels that are as yet little understood. What Crick and Watson had done was to discover the physical substrate on which heritable information is represented, much like silicon is the physical substrate of present-day computers. For both substrates it is impressive how the exacting requirements imposed on them can be achieved with as much miniaturization and economy as they are. However, no one would say that the secret of computers is in the silicon, since computers can be equally well realized in many other physical substrates, though perhaps not quite so economically at present. Indeed, one reason that computer development has been as rapid as it has is

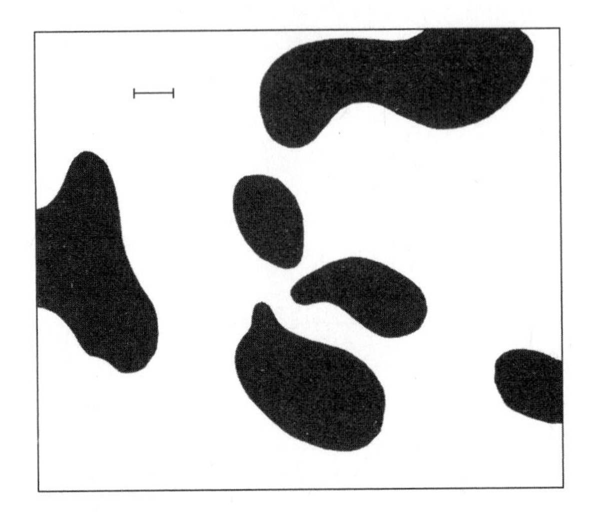

Figure 4.2 A dappled pattern reminiscent of animal fur derived by Turing by means of a computational process.[2] The particular pattern obtained is determined by minor random variations made in early stages of the process, rather than by any preprogrammed or genetic element. The short horizontal line is a scale indicator of the generating process. One gets very different but equally natural-looking patterns every time one runs the randomized process. Turing comments that he obtained this diagram in a "few hours by a manual computation"—evidently he did not have a machine available. (Copyright © 1952, The Royal Society)

that computer scientists made a conceptual separation at the very beginning between the physical technology in which the computer was implemented and the algorithmic content of what was being executed on the machines. This enabled hardware, software, and algorithms to evolve independently, and at their own spectacular rates.

Making similar headway in our study of biology, whether evolutionary or cognitive, demands the same separation of algorithm and substrate.[3] The distinction made here between a physical object and the information processing it performs is self-evident for anyone dealing with computers. The distinction is in no way subtle. Even for a traffic light one can easily distinguish between its symbolic function and its physical construction. But perhaps these distinctions were not quite so obvious in former times. The mind-body problem of Descartes and his followers may have been an earlier reference to such a distinction. But now when computers are ubiquitous, there is no reason for confusing "what it does" and "what does it."

Before moving on to the question of which algorithms might be realized by biology, I note that learning theory may inform the investigation of biology in a different sense also. A biologist performing experiments can be regarded as a learner who wishes to uncover the complex mechanisms of a particular system. As we shall see, there are inherent limits to the complexity of mechanisms that can be learned. The limits on what is learnable, which we will see in the next chapter, may be viewed as warning signals that the accumulation of experimental behavioral data by itself may not necessarily lead to progress in understanding how a system works. Individual human behaviors have been closely observed and widely recorded for thousands of years, yet we understand little about the mechanisms of the brain that gave rise to these behaviors.

After Alan Turing died, Max Newman, his mentor and friend, described in an obituary the central theme that had inspired Turing's many contributions to science: "The varied titles of Turing's published work disguise its unity of purpose. The central problem with which he started, and to which he constantly returned, is the extent and the limitations of mechanistic explanations of nature."[4] This is an insightful characterization, which we owe to someone who had known Turing well. It emphasizes the need for studying "extent and limitations," both of which were to become fundamental characteristics of computer science. The characterization further asserts the novelty of Turing's quest in suggesting that, while the established sciences—physics, chemistry, and biology—also aim for mechanistic explanations, nature also requires explanations of a kind that these older sciences do not address. In Turing's mid-twentieth-century writings one can already detect the pulse of the twenty-first century. Turing's place in history is assured by his discovery and successful pursuit of this previously unsuspected dimension to science.

Chapter Five

The Learnable

How can one draw general lessons from particular experiences?

All generalizations are false, including this one.
MARK TWAIN

5.1 Cognition

The idea that biological and cognitive processes should be viewed as computations appeared almost immediately upon the discovery of universal computation, and it was discussed by the early pioneers, including Turing and von Neumann. Because of subsequent slow progress in making this connection concrete or useful, some have despaired that it can never be made into more than metaphor, and that for fundamental reasons it cannot be made into a science. I disagree. I believe that developing any new science is fraught with challenges, and that we are making progress in this area at about the pace that might be reasonable to expect.

The universality of computation is what justifies this approach to cognition. Some have complained that the favored metaphor for the brain in every age has been the most complicated mechanism known at the time. Since the computer is currently that most complex mechanism, is it not a fallacy to adopt that metaphor? I would argue that the computer analogy goes beyond the fact that the computer is another complicated mechanism. What makes it different this time is the widely agreed universality of computation over all processes that we regard as mechanistic.

Turing and von Neumann both shared a second crucial insight: that mathematical logic, from which computation theory had emerged, was not the right grounding for a computational description of either thinking or life. In particular, Turing has the following memorable conclusion to his paper describing noncomputability results in logic: "The results which have been described in this article are mainly of a negative character, setting certain bounds to what we can hope to achieve purely by reasoning. These, and some other results of mathematical logic, may be regarded as going some way toward a demonstration, within mathematics itself, of the inadequacy of 'reason' unsupported by common sense."[1] This passage may be the first occurrence in science of the idea that common sense is somehow superior to reason. It foreshadows ample computational experience in the years that followed. While computers are extremely good at reasoning using mathematical logic, they find common sense much more challenging.

We are faced with two issues as a result: identifying what it is about common sense that logic fails to capture, and whether there is a scientific road to the problem of common sense. The first issue, I argue, is a result of mathematical logic requiring a theoryful world in which to function well. Common sense corresponds to a capability of making good predictive decisions in the realm of the theoryless. To address the second issue we need therefore a theory of the general nature of the theoryless. As I shall argue, the road we must take in that direction is paved with ecorithms.

The algorithms studied most widely in computer science aim to solve instances of some specific problem, such as integer multiplication or the Traveling Salesman Problem. These algorithms, by design, already incorporate the expertise needed for solving them. Ecorithms are also algorithms, but they have an important additional nature. They use generic learning techniques to acquire knowledge from their environment so that they can perform effectively in the environment from which they have learned. They achieve this effectiveness not by intensive design, but by making use of knowledge they have learned. The designed-in expertise is limited to generic learning capabilities, and their use. Understanding ecorithms requires developments beyond basic algorithmic theory. One now needs to analyze not only the algorithm itself but also the algorithm's relationship with its environment.

The theory of probably approximately correct, or PAC, learning deals with this relationship between the algorithm and its environment. It addresses the fundamental question of how a limited entity can cope in a world that in comparison is limitless, and does so while keeping to an absolute minimum any assumptions about that limitless world.

5.2 The Problem of Induction

Living organisms from the lowliest have some capability to adapt. They learn to avoid doing actions that are detrimental to themselves in favor of those that are beneficial.

In real-world environments an almost limitless number of distinct possible situations may occur. A useful learning capability therefore always needs to provide a significant component of generalization; a learned behavior has to be effective not only in situations that are identical to ones previously experienced but also in any number of novel ones. For this reason I identify generalization as the core of the learning phenomenon. Remembering a list of a hundred words shown once may be a challenge for us humans, but this is best regarded as a bug of our neurobiology, the legacy hardware architecture our species inherited. Because our brains lack the means of manipulating memory addresses in the way computers are able to do, and because each neuron is connected to only a small fraction of the others, memorization is unnecessarily difficult.[2] However, we humans are excellent at generalizing, a skill that is both philosophically fraught and difficult to endow in our computers.

There is a difficulty in placing generalization at the core of learning, at least for philosophers, who have argued for millennia that it is difficult to make a logical argument for rationally inferring anything from one situation to another that one has never before experienced. This is known as the problem of induction. Aristotle said that there are two forms of argument, syllogistic and inductive.[3] Here I interpret these words to mean that if one has a certain belief, then the belief was arrived at either by logical deduction (syllogism) from things already believed, or by induction (generalization) from particular experiences. In this formulation it is induction that is the more basic since it enables primary beliefs, whereas logical deduction requires some previous beliefs.

The main paradox of induction is the apparent contradiction between the following two of its facets. On the one hand, if no assumptions are made about the world, then clearly induction cannot be justified, because the world could conceivably be adversarial enough to ensure that the future is exactly the opposite of whatever prediction has just been made. This skeptical position is ancient. For example, the philosopher Sextus Empiricus wrote some 1,800 years ago:

> [The dogmatists] claim that the universal is established from the particulars by means of induction. If this is so, they will effect it by reviewing either all the particulars or only some of them. But if they review only some, their induction will be unreliable, since it is possible that some of the particulars omitted in the induction may contradict the universal. If, on the other hand, their review is to include all the particulars, theirs will be an impossible task, because particulars are infinite and indefinite. Thus it turns out, I think, that induction, viewed from both ways, rests on a shaky foundation.[4]

On the other hand, and in apparent contradiction to this argument, successful induction abounds all around us. Generation after generation, millions of children learn everyday concepts, such as dogs and cats, chairs and tables, after seeing examples of them, rather than precise definitions. Each child will typically see few examples of each concept, and the examples different children see will in general be different. Nevertheless, when asked to categorize a new example as to whether it is a cat or a dog, children will agree with each other on a high percentage of occasions, perhaps surprisingly high given the paucity and variability of the information they have been provided. From this we have to conclude that generalization or induction is a pervasive phenomenon exhibited by children. It is as routine and reproducible a phenomenon as objects falling under gravity. It is reasonable to expect a quantitative scientific explanation of this highly reproducible phenomenon.

While these two facets, the difficulty of justifying induction without assumptions, on the one hand, and the pervasiveness of induction, on the other, are on the surface contradictory, they are not implacably inconsistent. There may exist some acceptable assumptions that hold for the reproducible, naturally occurring form of induction, and under which induction is rigorously justifiable. I argue that this is exactly the case, and that just two assumptions are sufficient to give a quantitatively compelling account of induction. Further, these two particular assumptions are also necessary and unavoidable.

The first assumption is the Invariance Assumption: The context in which the generalization is to be applied cannot be fundamentally different from that in which it was made. If I move from one city to another, then I can benefit from my previous experience only on the assumption that things are not too different in the two cities. To put it a bit more mathematically, this assumption requires that the functional relationships and the probability distribution *D* that characterizes how frequently different situations arise remain somehow constant over time. It is important to note that the Invariance Assumption does not require that the world not change at all. It requires only that there are some regularities that remain true. These regularities may even specify how the world tends to change with time: If we observe the Sun going down toward the horizon in an interval of an hour, we expect that in the next hour the sun will go down even closer to the horizon, rather than that it will repeat the previous positions in a zig-zag fashion. Or, as the Wall Street financier J. P. Morgan, on being asked by a questioner for a prediction about the future course of the stock market, said: "It will fluctuate."

The second assumption is the Learnable Regularity Assumption. We are quite good, but possibly not perfect, at categorizing. If we look into an aquarium, we can fairly reliably distinguish between plants and animals, even species we have not seen before. We must be doing this by applying some criterion that distinguishes animals from plants. These criteria can be viewed as regularities in the world. Such regularities have been discussed as such by philosophers, notably by David Hume in the eighteenth century. Computer science adds at least two further levels to this discussion. First, it is essential to require that any useful criterion or regularity be detectable: Whether the criterion applies to an instance should be resolvable by a feasible computation. For example, the number of measurements we need to make on the object in question, and the number of operations we need to perform on the measurements to test whether the criterion of being an animal holds or not, should be polynomially bounded. A criterion that cannot be applied in practice is not useful.

However, the induction phenomenon has a second, even more severe, further constraint on it. It is not sufficient that the regularity or criterion just exist or even that it is detectable. To explain induction it is also necessary to explain how an individual can acquire the detection algorithm for the regularity in the first place. In particular, this acquisition must be feasible, requiring only realistic resources and only a modest number of interactions with the world. Of course, different kinds of regularity may require different levels of learning effort. Think about the night sky. Data about the positions of the visible objects has been available to our ancestors from the beginning, there for anyone to see. A little systematic observation revealed the easy-to-learn regularity that all the objects are in fixed positions relative to each other, except the few we call planets. It took thousands of years before someone, namely Kepler, discovered the much-more-difficult-to-discover regularity that the planets move in ellipses.

The Invariance and Learnable Regularity assumptions may seem restricting, but in fact they are liberating: They free the learner from certain responsibilities that are impossible to realize. The Invariance Assumption requires only that predictions hold for examples drawn from the same source as the examples were drawn during learning. If we learn from naturally occurring examples, then we only need to make good predictions about other natural examples. In the case of learning to distinguish animals from plants, this would imply, for example, that accurate predictions on artificial or mythical cases are not required. Computer-generated images of fictitious hybrids, designed to split human opinion exactly fifty-fifty as to whether they are plant or animal, will not be relevant to our interpretation of the induction phenomenon.

The Learnable Regularity Assumption also imposes some liberating limitations. It requires that some regularity exists, and that this regularity be effectively detectable for any example. It goes further in insisting that this regularity be learnable with moderate effort. A case that therefore need not be encompassed is where the examples are natural but then encrypted by some method that cannot be efficiently reversed. Thus the pictures of the animals or plants can be encoded so that they cannot be deciphered by any efficient computational process. This does not remove the regularity from the data if the original data is still recoverable in principle, even if only by an infeasibly laborious computational process. But if it is not practical at all to discover the regularity, then the regularity is no longer a *learnable* regularity, and it need not be addressed.

5.3 Induction in an Urn

I will show that these two minimal assumptions—the Invariance and Learnable Regularity assumptions—enable us to explain the possibility of induction rigorously. PAC learning is based on these two assumptions of invariance and learnable regularity. The next several sections will develop this idea in more depth.

Suppose you are presented with an urn containing millions of marbles, each one with a number written on it. You can reach in and draw a marble at random from the urn. You are allowed to draw 100 marbles. Your task is to determine which numbers occur at least once among all the millions of marbles in the urn. Is this possible?

The answer is clearly "no" if no assumptions are made at all, since it is possible that all the marbles have different numbers written on them. Any 100 draws will then fail to identify the numbers on the remaining millions of marbles. On the other hand, the answer is clearly "yes" under certain extreme assumptions. For example, if it is known that all the marbles are identical, then a single draw would give complete knowledge about all the marbles.

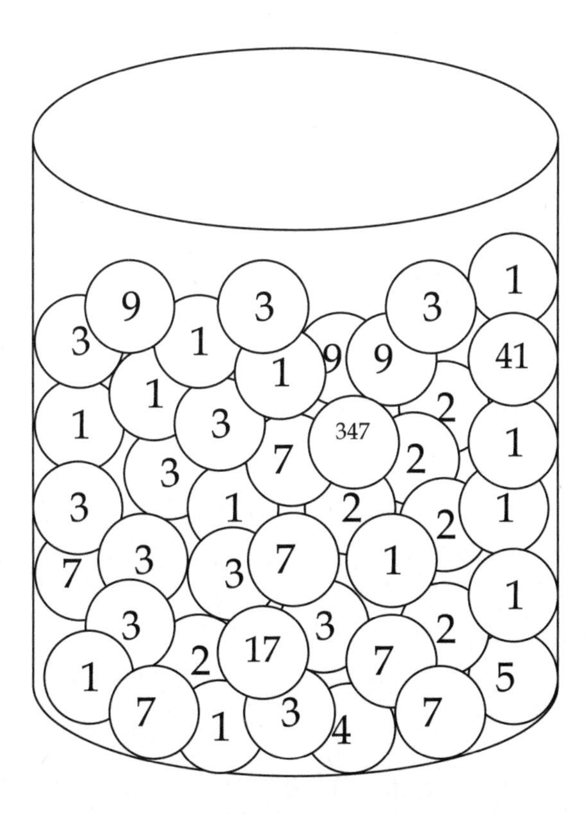

Figure 5.1 Given a large urn containing millions of marbles, can one induce which marble types are in the urn from a small sample drawn at random?

What is a little more striking is that there is a qualified positive answer under some more interesting intermediate conditions. In particular, if the number of different kinds of marble in the urn is not one but a small number, such as five, then one can still achieve a useful level of generalization.

The reason why having a fixed number of marble types makes some level of induction possible is simply the following. Any marble that occurs frequently enough will be among the 100 drawn with high probability, and hence the 100 marbles drawn will be representative of the most frequently occurring marbles, unless, of course, you had particularly bad luck with the draw.

To argue this more precisely, first consider the specific case that as many as 50 percent of the marbles are of one type. Suppose this 50 percent all have a 3 written on them. Then the probability that all 100 random draws miss a 3 is $(1/2)^{100} = 0.00000000000000000000000000000078886\ldots$ This is the same as the probability that 100 tosses of a fair coin all come up tails. The likelihood of this occurring is, of course, extremely small—so small, in fact, that if an experiment of tossing 100 coins had been repeated every nanosecond since the currently estimated date of the Big Bang, then the probability that this eventuality would have ever happened is still less than 1 in 2,000. Therefore, for the urn in which 50 percent of the marbles are 3s, we are safe to conclude that after 100 draws, with overwhelming probability, we will have seen at least one 3, and hence a representative of at least 50 percent of the contents of the urn.

Such a very specific assumption, that 50 percent of the marbles are the same, is not essential either. The knowledge that the number of different marble types is small also gives a sufficient principled basis for induction. Suppose that we know that there are at most five different marble types, but have no idea of their relative frequency. It turns out that after 100 picks we should have high confidence that we have seen representatives of over 80 percent of the contents of the urn. The argument goes like this. If any of the five marble types occurs with frequency more than 5 percent, then the probability that that type was missed all 100 times is less than $(1 - 0.05)^{100} = (0.95)^{100} < 0.6$ percent. Since there are at most five such types, the probability that any one of these frequent ones has been missed is less than five times this quantity, namely 3 percent. There can be at most four types that occur with frequency less than 5 percent each, and hence the rare types account for less than 20 percent of all the marbles between them. Combining the two sources of error, we conclude that with probability at least 97 percent no marble type that occurs with probability greater than 5 percent has been missed, in which case the missing marble types can account for at most 20 percent of the contents of the urn. Hence you will have drawn representatives of at least 80 percent of the marbles, unless you are unlucky and miss some common types. But you are unlucky in this way less than 3 percent of the time.

In other words, you can predict with 97 percent confidence that after 100 picks you will have seen representatives of at least 80 percent of the contents of the urn. This is in spite of the fact that the distribution of the various types of marbles is arbitrary and unknown to you. It may be that each of the five occurs with equal 20 percent probability. Or it may be that 92 percent are of one kind and the other four kinds occur with frequency 2 percent each. Or it may be that three of the marble types are extremely rare, each occurring 0.1 percent of the time, and the remaining two each occurs 49.85 percent of the time. The claim is equally valid for these three cases, as for any other.

All we need to make this claim are our two assumptions: that the contents of the urn do not change (invariance), and that there are a fixed number of marble types represented in the urn (which provides a sufficient learnable regularity). From this it was possible to deduce totally rigorously that from a small sample one can make meaningful predictions about future draws from the urn.

5.4 Error Control

Clearly, then, induction with our minimal assumptions is powerful. Equally clear is that we cannot avoid errors, which can come from two sources. The first source is that of rarity errors. There may exist in the urn some rare types of marbles that are unlikely to be drawn in any small sample. Their existence therefore will be predicted with correspondingly small probability. The second source is that of misfortune errors. With some small probability the sample drawn will be unrepresentative of the contents of the urn because it missed some common marble types. In extreme enough cases of such misfortune, as when all the common marble types are missed, the sample will not support any useful claims about what is in the urn.

The interesting thing is that while neither of these two sources of error can be totally eliminated, both can be controlled. By this I mean that the probability of both sources of error can be driven down to arbitrarily small (but nonzero) quantities by increasing the number of marbles drawn. Most significantly, controlling error is affordable: The cost in increased sample size that needs to be paid will depend only modestly, in fact polynomially, on the predictive error that is to be tolerated. That affordability is crucial—if good predictions could be achieved only in an idealistic limit requiring exponentially much effort, then explanations of real-world phenomena would not follow.

In this urn model the only action is to pick marbles and there is no other computation. Hence the computational cost may be regarded as the number of marbles picked from the urn. I shall call this number S. By feasible I have a concrete quantitative notion in mind. I want S to increase only polynomially both with n, the number of different marble types in the urn, and with $1/error$, the inverse of the maximum error one is choosing to tolerate. Here *error* will be either the rarity error or the misfortune error, whichever is smaller. An instance of a polynomial bound for the error is the quadratic bound $(1/error)^2$. Then if one is willing to tolerate 10 percent errors (i.e., $error = 0.1$), then the number of examples and total number of steps needed to achieve that level of accuracy should be proportional to $(1/error)^2 = 10^2 = 100$.

We can find out just how affordable our predictions can be. If we assume, for example, that all n types occur in the urn with the same probability, to draw a set representing half the available types, we will need to choose at least $n/2$. In this case the cost of a good induction will be proportional to the number of available types. It can also be shown that for some distributions the dependence on the inverse error can be similarly linear. This shows that some minimum price may have to be paid for good generalization. Fortunately, it can be shown that for no distribution is the actual price ever much higher than these bounds, which are linear in n and in $1/error$.[5] The argument needed to prove this is a generalization of the one just given for the particular case of five marble types, where it was shown that a sample of 100 marbles was sufficient to get a confidence of 97 percent of having a prediction error less than 20 percent.

5.5 Toward PAC Learnability

The urn example establishes the feasibility of induction in that setting, but it is only a restricted setting; the number of distinguishable objects is small, in fact small enough that it is practicable for the learner to witness a big fraction of them. We have already seen Sextus Empiricus's objections to such an assumption. In the real world the number of distinguishable objects is so large that no learner can expect to see more than a minute fraction, even in a lifetime. A child will have seen only a small fraction of the millions of distinguishable individual animals and yet be able to classify examples according to species. The onerous requirement on human induction, which the urn example does not satisfy, is therefore that after seeing S examples, it has to be able to generalize over sets that have far more than S distinguishable individuals.

Hence the question we have to ask is this: What kind of induction is feasible that matches the urn example in rigor and practical feasibility, but can induce over sets of as many as, say, 2^S distinguishable types, rather than merely S, but with a cost that is polynomial in S, and not exponential?

Fortunately we can adapt the urn example for learning in such a more expressive and exacting setting. Suppose, for simplicity, that the features we detect when trying to distinguish an animal from a plant on a certain planet all have just yes/no values. Suppose also that there are just twenty such features, including the following: is grey, is red, is green, is brown, is small, is big, has eyes, has legs, has leaves, has long ears, can move, can breathe, and so on. Assuming a criterion in terms of these particular features exists, then for each possible combination of yes/no values of the twenty features that applies to some creature it is completely determined whether that creature is an animal or a plant.

The problem is that twenty features, each being present or absent, can be combined in $2^{20} = 1{,}048{,}576$ ways. Now if we have the opportunity to observe many millions of creatures, each identified as animal or plant, we are essentially back to the urn model. We will see all commonly occurring combinations of features, view each different combination as a type of marble, and will then be able to classify future configurations with the same confidence that the calculation for the urn model justifies. However, a scenario where it is necessary to see exponentially many examples in terms of the natural parameter, here the number of features, is simply unrealistic. Even if not all the 2^{20} different combinations of features occur in nature, a large number may, perhaps in the thousands. Humans can learn from far fewer examples even in cases such as this where the number of distinguishable individuals is really enormous. Our algorithm should be able to, as well.

This brings me to the definition of learnability, or the requirement that we can reasonably impose on a learning algorithm before we declare it to be successful. In this definition we shall demand that an algorithm should learn from a number of examples that is polynomial in the number of the features n. The need for n or n^2 or even n^3 examples may be acceptable, but exponentially many, such as 2^n, would not. (The number of distinguishable types, however, may be as many as 2^n.) We also want to control the error, and insist that this control be again polynomial. In order to achieve all this, we therefore insist that the computational cost of the process of deriving the induced generalization from the examples is polynomial not only in n and but also in 1/*error*. Note that this computational complexity criterion already implies the polynomial limitation on the number of examples drawn, or the sample complexity, since it takes at least one operation to process each example.

The next question to ask is whether this notion of induction is not so onerous that it is unachievable. We can show that this is not the case—induction in the sense of this definition can be attained for certain useful classes of concepts. One such class is that of conjunctions. A conjunction is an expression that specifies for each feature whether it must hold, it must not hold, or it does not matter whether it holds. An example of a conjunction in the present case is

(can move = **true**) **and** (has eyes = **true**) **and** (is green = **false**).

This expresses the criterion that "can move" and "has eyes" must hold, "is green" must not hold, and the remaining seventeen feature values do not matter. Such a conjunction, in turn, defines the concept of an animal on a certain planet if and only if every animal there satisfies the conjunction (i.e., satisfies all three components) and every nonanimal fails to satisfy it (i.e., fails to satisfy at least one of the three components.)

Now let us assume, for the sake of argument, that the concept we are trying to learn can be expressed by exactly this conjunction. In other words, we are assuming that everything on that planet that can move and has eyes and is not green is an animal, while anything that fails to have at least one of these three properties is not. How would we learn the conjunction efficiently from a modest number of examples? It turns out that the marbles and urn analysis of the previous section can be adapted to apply here also, but with the polynomial bound now in terms of the number of features n, rather than only in terms of the number of distinguishable types, which may be 2^n or exponential in that quantity.

To see this, we can treat each example of an animal or nonanimal as a marble that has written on it for each of the twenty features whether the feature holds for the example or not. We only put marbles corresponding to positive examples of animals in the urn. The marbles representing animals will all be labeled by the statement: can move = **true**, has eyes = **true,** is green = **false**, while the other seventeen features between them can take on any of the 2^{17} different combinations of the remaining feature values in any arbitrary ratios of relative frequency.

Our learning algorithm works like this. It forms a list of the $2n$ possible properties, for each feature one property asserts that the feature is true and the other that it is false. We call this list L. The list L initially contains all $2n$ properties, or forty in our running example. Marbles are then drawn one by one. If a marble is drawn that misses some properties that are in L, those properties will be deleted from L, because they are properties that not all animals share. For example, if one animal is not grey, then greyness cannot be a necessary property for all animals, and this property should not be in the conjunction. After 100 marbles have been drawn, the conjunction of the properties remaining in L is declared to be the hypothesis or criterion for animals.

This procedure, known as the elimination algorithm, induces an accurate criterion for recognizing whether something is an animal. The reason is the following. First, all the properties in the correct conjunction for animals must be present in the final L since every marble in the urn had all the properties that all animals share, and the only properties deleted from L were those that were missing in at least one animal. So the only possible source of error is that some property, such as "has long ears," remained in this final L, while it should not have. This would mean that in 100 trials every animal drawn had long ears. If this property is not essential to animals, then the ones that *lack* this property must be truly rare (rarity error) or we were unlucky in our pick of 100 animals (misfortune error). Exactly as in the urn argument, we can argue here also that, with high probability, the properties that falsely remain in L (e.g., having long ears) must be those that between them exclude no more than a small percentage of animals. In fact, essentially the same polynomial bound can be proved in terms of n (the number of features, not animals) and $1/error$ as for the urn problem.[6]

That correctly classifying animals via conjunctions is no more difficult than the urn problem is perhaps surprising, since there were just n different marble types in the first scenario, while now there are 2^n different types of animals. Let us therefore reexamine our assumptions.

First, we made the Invariance Assumption that the examples encountered in the testing phase come from the same source as in the learning phase. Examples rarely seen during learning will be equally rare during testing, and therefore less important for the learner to know about.

Second, we made the Learnable Regularity Assumption. In this case we assumed that for the given features a criterion for distinguishing animals from plants could be expressed as a conjunction. This was sufficient because conjunctions can be shown to be learnable, as we have just seen.

As we shall see later, many function classes, because they seem not to have learnable regularity, appear not to be learnable even when the Invariance Assumption holds. In other words, the Learnable Regularity Assumption substantively constrains what learning algorithms can do. That may seem a problem, but in fact such constraints are needed to make learning possible. The fact that the elimination algorithm for conjunctions used only positive examples can help us see why! It may seem impossible to learn to classify animals and plants by looking at only animals. Nevertheless, as we have seen, it is both possible and even rigorously justifiable. The reason is that the constraint that there exists a conjunction that distinguishes one type from another is in itself highly informative. In this case it permits learning from positive examples alone. (As a very loose analogy for how information can be conveyed by constraints, suppose I challenged you to solve a puzzle that I claim to have solved myself. In your search for a solution you would find it helpful to know whether it had taken me ten seconds, ten hours, or ten days, even though this information sheds little light specific to the problem.)

Equally learnable are disjunctions, which are expressions of exactly the same form as conjunctions, except that each **and** is replaced by an **or**, requiring only that at least one of the listed set of the properties holds, rather than that every one of them holds. We would then assume that the concept of an animal is expressible as a disjunction such as x_3 **or** x_7 **or** x'_9, where x'_9 denotes **not** x_9. Like conjunctions, disjunctions are learnable, albeit by relying entirely on negative examples, and again using the elimination algorithm. So, if the algorithm encounters a brown plant, "is brown" gets removed from the list of traits that each guarantee something being an animal.

Eliminating what has to be eliminated, as we do here, is, of course, a long-recognized principle of reasoning. Francis Bacon, in the early seventeenth century, and John Stuart Mill, in the nineteenth, both emphasized its importance.[7] Sherlock Holmes was even more categorical: "When you have eliminated the impossible, whatever remains, however improbable, must be the truth. " Unfortunately for Mr. Holmes, exploiting the elimination method in a foolproof way is rarely practicable. Some cases will remain on the list that are not true. The PAC framework offers the needed analysis of the error that will result.

It is natural to ask whether conjunctions and disjunctions are expressive enough to account for human concepts. One way of phrasing this is to let the features be words in a certain dictionary, and ask whether each word in the dictionary can be expressed in terms of, say, a conjunction of the others. The philosopher Ludwig Wittgenstein argued that the notion of a game has no feature that is common to all instances of it. For example, not every game is won or lost, or played by two people, and so on. This implies that conjunctions are not enough for expressing everyday words in terms of each other, since such conjunctions would have to contain exactly the features that are essential to all instances. A similar argument can be made for disjunctions.

Circuits that consist of **and** and **or** statements, composed in an arbitrary way, rather than in a single layer as in conjunctions or disjunctions, are much more expressive. If one has to learn such a circuit, and intermediate nodes in it do not correspond to natural concepts for which labeled examples are available, then the ability to learn conjunctions and disjunctions is not enough. Indeed, as we shall see later, it is widely believed that some function classes that are more expressive than conjunctions or disjunctions are *not* learnable. The questions of determining the most expressive classes of functions for which learning is still possible are the most fundamental questions of learning theory.

5.6 PAC Learnability

What we have been describing is a notion of probably approximately correct (PAC) learning. When first introduced, the corresponding class was simply called learnable, in analogy with Turing's notion of computable, to indicate that what was being sought was a robust characterization of what was practically learnable by explicit computational means. The "probably" acknowledges misfortune errors, and the "approximately" rarity errors.

Many of the concept classes we have seen so far are PAC learnable: conjunctions and disjunctions are, as well as the class of linear separators discussed in Chapter 3. The perceptron algorithm described there is not quite sufficient to establish this. One impediment is that the number of iterations of the perceptron algorithm will be exponential if some examples are exponentially close to the separator. Fortunately, this impediment can be overcome by using a different algorithm, one based on linear programming.

The critical idea in PAC learning is that both statistical as well as computational phenomena are acknowledged, and both are quantified. There have been earlier attempts to model induction using purely computational or purely statistical notions.[8] However, I believe that combining the computational *and* the statistical provides the key to understanding the rich variety of learning phenomena that exist. The notion of PAC learning is concerned with describing what needs to be achieved in order to constitute induction. It is neutral on both which concept class should be learned and which algorithm is to be used to learn it. What it does do is to offer a quantitative analysis of learning. Which algorithms the human nervous system uses and which classes are being learned are not currently known. But at least we have a way of making these questions concrete.

By now it may have occurred to the reader that the model described is undoubtedly a simplification of the broad range of phenomena that humans manifest in relation to learning. In itself the model addresses the core phenomenon of computationally feasible induction from examples. The model can be and has been extended in numerous directions so as to capture many additional aspects of learning.[9] These directions include allowing for some kind of noise in the data, the concepts changing slowly rather than staying invariant, the learner asking certain questions, or the algorithm working only for specific distributions. Having a definite mathematical model of the inductive process gives a vantage point for investigating these important facets of learning.

5.7 Occam: When to Trust a Hypothesis

The great advantage of a hypothesis generated by a PAC learning algorithm is that it comes with a reliability guarantee. However, it is often the case that we are confronted with a hypothesis about the provenance of which we know nothing. A hedge fund manager, for example, might be told that some specific pattern of price fluctuation has been present in market activity for some time. He must decide if he should start using this pattern to make investment decisions, even if no information is available about how the pattern was identified or who had identified it. This kind of question also arises routinely in the practice of machine learning, where many algorithms are employed that have not been proved to be PAC learning algorithms but are useful nevertheless. Happily there are some entirely rigorous criteria to apply to such situations also.

The answer lies with Occam algorithms: They provide a rigorous approach, even in such cases of total ignorance about the origins of a hypothesis, and exemplify the role of purely statistical arguments in machine learning.[10] What this approach provides are some conditions under which an unfamiliar hypothesis can be trusted. These conditions make concrete and rigorous the intuition sometimes attributed to the fourteenth-century logician William of Ockham that all things being equal, simpler hypotheses are more likely to be valid than complex ones.

Suppose that you are trying to predict horse races, and that someone gives you data from a hundred past races in which every time the heaviest horse won. Discerning whether the heaviest horse is the sure thing it might appear to be requires several steps. First, you will need to be convinced that the 100 races you are shown were not maliciously selected from among many that overall showed no such clear pattern. Second, you would need further reassurance that the data was not from a different planet. These two requirements are roughly equivalent to the Invariance Assumption of the definition of learnability, that future events will be drawn independently from the same probability distribution as the 100 events in the dataset. Third, you would need to assess the complexity of the hypothesis. It is tempting to bet on the heaviest horse because of the simplicity of the rule "the heaviest will win." It seems unlikely that 100 races would all satisfy such a simple rule just by accident. If the rule were much more complex, for example that the horse's height, the owner's weight, and the trainer's age (all in appropriate units) added up to a prime number, then you would be a little more skeptical, and justifiably so. Even if the winners were totally unpredictable and arbitrary, some prediction rule could always be engineered to match them after the fact, if the rule is allowed to be complicated enough.

This intuition can be made rigorous as follows. Suppose that you have a well-defined language for expressing hypotheses, and suppose that at most N distinct hypotheses can be expressed in this language. Suppose also that someone gives you a dataset of S examples that all agree with h, one of the N permitted hypotheses. To decide whether to accept this hypothesis h, we need only the Invariance Assumption. Suppose that a fixed rule h^* is bad, in the sense that it predicts correctly only a fraction p of the examples from the distribution. Then the probability that it will be accurate on every one of the S independently drawn examples will be p^S. This will be an extremely small fraction if S is large and p is substantially less than 1. For example, if $S = 100$, and p is less than 0.8, then the probability that this bad rule will predict correctly every one of 100 random examples is about 0.0000000002, or 2×10^{-10}. This is smaller than the probability that thirty-two successive tosses of a fair coin will all come up heads, an extremely unlikely event.

This argument applies to any one fixed rule. But what if an adversary tried to trick us by choosing the rule to fit the data after having looked at the data set? We know there are certain limits to what the trickster can do. There are a limited number N of hypotheses from which he can choose; for example, if the hypotheses are conjunctions over n variables then $N = 3^n$, since each of the variables has three options, present in the conjunction, present in negated form, or absent. In particular, if the probability that any one bad hypothesis looks good is no more than 2×10^{-10}, then the probability that at least one of the up to N bad hypotheses looks good is no more than N times this, or $N \times 2 \times 10^{-10}$. Even if the trickster had a million different rules to choose from, the odds of him finding one that would be both bad and an effective trick is only 1 in 5,000. In fact, as long as N is much smaller than 5 billion, we can be confident that there will be no rule among the N that classifies all the 100 examples correctly but is in fact less than 80 percent accurate on the distribution D in general.

So, as an example, imagine race horses are coded according to a list of 1,000 traits, and that each hypothesis is a conjunction of three traits, such as "largest **and** darkest **and** oldest." There will be about 166 million distinct hypotheses. (This is because if one makes a sequence of three choices, each with a thousand outcomes, there will be 1 billion outcomes overall. Because the order of the traits, if distinct, doesn't matter, only about one in six such hypotheses is unique.) Hence the probability that even one bad hypothesis among them agrees with all the 100 examples will be slightly more than .03. So if "largest **and** darkest **and** oldest" correctly predicted the winner of 100 randomly chosen races, you would have to be irrational not to bet that way on the next one. We can make similar calculations, and determine what level of confidence in a given rule is justified, even with fewer than 100 examples, and even when the rule predicts not all but only most examples correctly.

What we have shown is that we can depend on the predictive power of a rule someone has given us if we are sure of three conditions. First, we need to be given a data set of past examples that have high agreement with the rule. Second, we need to know that the rule is from some small class of rules that was fixed before the examples were selected. And third, we must have convinced ourselves that the examples presented to us were chosen randomly and independently from the same distribution from which we will wish to make future predictions—that is, only rules trained on thoroughbreds should be used to bet on thoroughbreds.

Now, what is the relationship between PAC learning and these Occam algorithms? In the case of PAC learning we have a guarantee, ahead of seeing any examples, that the learning algorithm we have in hand, such as the elimination algorithm for conjunctions, will yield a good predictor whenever the hidden function to be learned lies in a certain class of concepts. In the case of Occam algorithms, the PAC-like guarantee of predictive accuracy does not depend on the process by which the hypothesis was generated; it is provided only for one specific hypothesis at a time. But that can be liberating, as the Occam argument then gives us a rational justification for trying learning algorithms that do not always work. If we are lucky, and find a hypothesis that explains the data and is short enough to have predictive power, then we can go ahead and use that hypothesis for making predictions. If we are unlucky, and the hypothesis obtained is too long or does not fit the data sufficiently, then we will recognize this failure immediately and not use it to predict, so that no harm will have been done.

One might say that the most reliable way of testing a hypothesis is to do a test on new data—that is, data that has not been used in deriving the hypothesis. While this is true, it is not free of cost, since it requires that we retain some data that does not inform the hypothesis. Using that extra data, we might be able to produce a better hypothesis. When one does go live with a hypothesis—whether betting on a horse or recommending an investment or a medical treatment—one inevitably has to make an Occam-like decision: Given all the available data what exactly is the best hypothesis that can be deployed?

5.8 Are There Limits to Learnability?

We have seen a variety of learning algorithms. The perceptron algorithm is one; the elimination algorithm is another. Obtaining a succinct hypothesis and appealing to an Occam argument is a third approach. As we shall see in Chapter 9, many learning algorithms are now known and some are already widely deployed. Also, there are no doubt many more algorithms that no one has yet conceived. But where do we look to for ultimate limits? As pointed out earlier, learning is based on a deep interplay of computational and statistical phenomena. If there are limits to learning, then these are the directions in which they will be found.

First consider statistical limits, which, though weak, are significant. These impose a condition on the minimum number of training examples needed in order to learn reliably. This number does depend on the distribution. For easy distributions high accuracy can be reached with few examples. An extreme case of an easy distribution is one that has only one positive example that ever occurs, in which case seeing that once will give all the information that is available or needed. The bounds obtained for worst-case distributions often provide useful guidance on how many examples to use in practice. One can show that, for certain distributions of examples, the minimum number of examples required is proportional to the ratio of the number of variables n and the error to be tolerated.[11]

Computational limits are more severe. The definition of PAC learning requires that the learning process be a polynomial time computation—learning must be achievable with realistic computational resources. It turns out that only certain simple polynomial time computable classes, such as conjunctions and linear separators, are known to be learnable, and it is currently widely conjectured that most of the rest is not.

This intuition can be expressed more precisely with an Occam-style argument. Suppose that a function of n variables that describes some regularity is detectable or computable in, say, n^2 steps. (Note that conjunctions can be detected in about n steps, but some useful regularities may be more complex.) Then the behavior of the function on a little more than n^2 randomly chosen inputs will determine the behavior on most of the exponentially many possible inputs in a PAC sense. In other words, the hidden function will be largely determined, for any distribution D in question, by its values at a polynomial number of inputs. The reason that most such functions do not appear to be learnable with polynomial effort is not that in a polynomial amount of data the function is not already implicit, but that this implicit information cannot be extracted from that data with polynomial effort.

I believe that the primary stumbling block that prevents humans from being able to learn more complex concepts at a time than they can, is the computational difficulty of extracting regularities from moderate amounts of data, rather than the need for inordinate amounts of data. For example, the difficulty of discovering the elliptical nature of the orbits of the planets was not that the amount of data needed took hundreds of generations to compile, but that elliptical orbits as seen from Earth did not constitute a regularity that humans found easy to extract.

Yet another way of stating the relative roles in learning of computation and statistics is to observe that if the assertion that P = NP (or equivalently, that the NP-complete problems are computable in polynomial time) is true, then all of P would be PAC learnable. Recall that NP is the class corresponding to mental searches. If P is equal to NP, then one can take a polynomial number of random, labeled examples and then simulate in P a machine that realizes the necessary mental search for a hypothesis that agrees with these labeled examples. This would give a hypothesis with an Occam guarantee of good predictive accuracy on future examples. In other words, the truth of the computational assertion that P = NP would imply that all polynomial computable functions would be learnable.

We conclude from this that if we are to understand the limitations of learning, we need to look at computational limitations. Unless it turns out that P = NP, or some other unexpectedly strong enough positive result is true, the notion of knowledge being implicit in data is not sufficient to mark the boundaries of what is learnable.

So what are the computational limitations on learning?

We get a first clue from the study of cryptography. This field is concerned with designing algorithms for encoding a message so that an intended recipient can decode it, but any unwelcome eavesdropper who intercepts the message cannot. To make this decoding possible for the intended recipient, but not for the eavesdropper, the intended recipient must have as a key information that is not available to the eavesdropper.

In traditional cryptography the key is conveyed to the recipient by some secure process, such as inside an actual physically locked box or by a whisper in the ear from a trusted emissary. In public-key cryptography no such secure physical process is needed for transferring the key.[12] Instead, the recipient generates a "public key" and transfers it to the sender through public means, but the recipient privately retains some information, the "private key." In the RSA system the public key is the product of two large prime numbers p and q, and the private key consists of the prime numbers p and q themselves. Anyone can encrypt a message intended for the recipient with the public key, but only the recipient with the aid of the retained secret private key will be able to decode it in polynomial time. To ensure that the number of potential private keys is so large that it is not practicable for the eavesdropper to simply try them all in turn, the key should be, say, a thousand-digit number, of which there are too many to enumerate.

The conjectured infeasibility of computing p and q is believed to make RSA cryptography immune to attack. This conjectured immunity sheds light on why not all hypotheses in class P are learnable. One type of attack on any cryptographic scheme is known as chosen-plaintext attack. In this a would-be eavesdropper—perhaps an insider—is assumed to have access to the encryption device, and can feed the encryption device with many, possibly carefully chosen, pieces of text and then observe their encodings. From this information the eavesdropper extracts a decryption algorithm (equivalent to the key) that will decrypt any encoded message. In essence, then, a chosen-plaintext attack is similar to our learning scenario: the encoded messages are the examples, the bits of the original raw messages are their labels, and the decryption algorithm encapsulated by the private key is the concept to be learned. Assuming that RSA, or some other encryption scheme, is in fact resistant to chosen-plaintext attack, it follows that the class P is not

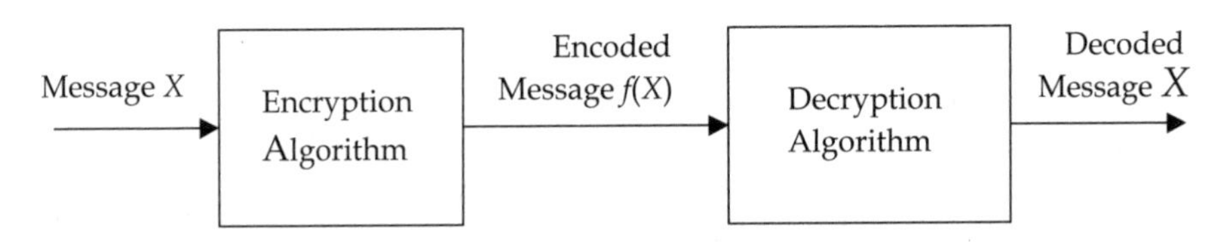

Figure 5.2 The existence of strong encryption methods implies the existence of functions that are computationally intractable to learn. The reason is that if all decryption algorithms were learnable from examples, then one could break any scheme by collecting enough pairs $(X, f(X))$ by feeding the encryption device with enough known messages X and intercepting their encodings $f(X)$. The pairs $(f(X), X)$ are then labeled examples for the decryption algorithm. Learning this algorithm amounts to breaking the code.

learnable, since if it were then we could learn all decryption functions and hence break all such cryptographic schemes.[13]

We do not need to look to just cryptography to find apparent impediments to learnability. Language gives us another domain.

In the 1950s the linguist Noam Chomsky considered various alternative classes of formal languages as the possible grammatical bases of natural languages, such as English.[14] The simplest in this so-called Chomsky hierarchy of formal languages are the regular languages. The mechanisms that generate them are finite-state automata, which are also among the simplest interesting computing mechanisms. An instance is illustrated in Figure 5.3.

Each such automaton can be represented as a diagram with a start node and a number of finish nodes. Each link in the diagram is labeled from a fixed set of symbols, such as *a* and *b*. The language accepted by the automaton is the set of all sequences that label paths in the diagram from the start node to some finish node.

Consider the finite automaton depicted in Figure 5.3. The language this accepts consists of all the sequences of *a*'s and *b*'s of length 5 that begin at the start node on the left and follow paths from left to right and terminate at the finish node. This automaton was generated by first drawing the lattice diagram shown, and then randomly labeling one of the two outgoing edges from each node with an *a* and the other with a *b*. Now, as you might observe, there are thirty-two possible sequences of five letters made from the set of *a* and *b*, but only sixteen distinct paths one can take from start to finish. Hence the language generated by this automaton only accepts one-half of the possible sequences as valid. This construction can be generalized so that if there are *l* letters (or other symbols, such as punctuation or numerals) in an alphabet,

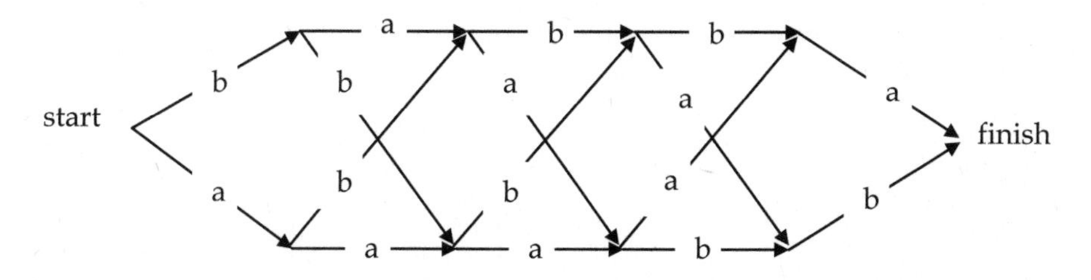

Figure 5.3 An example of a finite automaton of width 2 and depth 5. It accepts half of the sequences of *a*'s and *b*'s of length 5. For example, *bbbab* is accepted, but *aaaab* is not.

and if sentences are of length s, then there are l^s possible sequences and l^{s-1} of these, or a fraction $1/l$, would be accepted by the corresponding automaton.

If we could see all the l^s possible sequences and be told which ones are in the language and which are not, then we could predict new examples, simply because we have all the facts. This is just the urn model again. But this, of course, would be of exponential complexity in terms of the length of sentences.

The task of learning a regular language is that of taking a polynomial number (in terms of sentence length and alphabet size) of training sequences, each labeled according to whether it belongs to the language or not, and on that basis predicting for unseen sequences whether or not they are in the language. Inducing the hidden automaton or an approximation to it would be one possible approach. In principle, it is also possible that a learning algorithm would yield a hypothesis that can label new examples probably approximately correctly, but does not reveal an explicit description of the automaton. Several decades of research following Chomsky, however, failed to uncover such a learning algorithm. This failure was explained in the 1980s by a proof that any algorithm for PAC learning regular languages would imply a method for breaking the RSA cryptosystem.[15] Unless such systems are breakable, no computational learning algorithm for regular languages can exist.

The problem can be understood through the frame of the Occam argument. A random set of polynomially many examples is sufficient to essentially determine the hidden automaton. The difficulty, however, is that this determination is only in the implicit sense that automata that are very different will be most likely inconsistent with the sample. It is in extracting the automaton from this sample by a feasible computation where the difficulty lies—no way is known of doing this extraction in polynomial time. Language learning is of some consequence. We cannot hope to understand what human languages are without understanding how they are learned. A formal language that cannot be learned, cannot be the basis of human language.

All known algorithms for learning regular languages appear to work in exponential time as a function of the length of the sequences of symbols. Whether this can be improved, even for the special class of automata generated uniformly at random using the lattice diagrams as illustrated, is currently an open problem. Any reader not convinced that there are real computational impediments to generalization should take as a challenge this simpler goal of finding a PAC learning algorithm to categorize the output of a lattice automaton as words or not-words in polynomial time.

Finally, we note that there remain many natural classes of functions whose learnability is currently totally unresolved. For these it is not proven that they are PAC learnable, but equally there is no evidence that they are not. The prime example is the class of functions known as *disjunctive normal form*, often abbreviated as DNF. These are functions that can be expressed as polynomial-size expressions in terms of the number of variables, where the general form of the expression is an **or** statement joining subexpressions consisting of **and** statements of the variables, for example, (*X* **and** *Y*) **or** (*Z* **and** *T*). This can be viewed as a two-level representation, composed of conjunctions and disjunctions (which are themselves one-level representations). DNF is clearly more expressive than disjunctions and conjunctions separately. The best algorithm currently known for learning DNF has complexity exponential in the cube root of the expression size, which is still exponential but not of the worst kind and curiously similar to the best bound known for factoring integers.[16] DNF is an archetypal two-level representation, perhaps close to the boundaries of learnability. Resolving whether or not it can be PAC learned is a major open problem in learning theory.

5.9 Teaching and Learning

Learning a single concept from examples is already a striking natural phenomenon. But even more impressive is how humans can acquire expertise involving many complex interrelated concepts. Of course, humans often take many years to accomplish such feats. Our species invests up to two decades to educate its young, and there is no evidence that this is unnecessarily inefficient. But what view should we take of these more complex accomplishments?

In a college course one expects to learn more than just a set of concepts that could be learned equally well in any order. Rather, one expects to see a sequence of concepts, such that the later ones become accessible to the learner only after the earlier ones have been mastered. This is the paradigm of learning that we adopt here. To put it more mathematically, at any instant we are in a position to learn a new concept that is a member of a learnable concept class *C* with the set *X* of concepts that are already recognized as features. Once we have learned a new concept in this way, it becomes an added recognizable feature for the purposes of learning additional concepts in subsequent phases of learning. Each such new concept becomes in one's brain an equal citizen with all the features X that were available as features in previous phases, and gets added to X for the next phase. For example, the term "data" has a certain meaning in any time and place. Once we have some familiarity with that meaning, we can learn to recognize derivative concepts, such as that of "big data," that may have been out of reach before "data" had been learned and became a feature. A basic PAC learning algorithm operates up to the level of complexity that is expressible by members of C. At higher levels we can think of the teacher as fulfilling a part of the role of a programmer in defining a sequence of concepts that can be PAC learned in that sequence, but perhaps in no other.

This analogy between a teacher and a programmer also highlights some essential differences between the two. When programming a computer, the programmer needs to understand what exactly the existing programs already implemented on the machine do. This is not the case for a teacher, who does not know exactly the meaning to the learner of each word the teacher is uttering. It is possible for a teacher to get across the notion of "dog" by showing examples without knowing precisely which features of dogs are recognized by the learner, and how exactly these are interpreted. Much more specific knowledge is required of a computer programmer.

This disconnection between the teacher and learner is not entirely deleterious. It offers unique advantages to learning systems that programmed systems lack. A teacher can convey information by suggestions and examples, without knowing the exact state of the learner. In contrast, the programmer has to know the exact state of the system, including the exact functionality of the previously programmed features, if the new program is to work as intended. One can go further and say that the inherent strength of the teacher-learner relationship is that it works even when the exact state of the learner is not known explicitly by anyone, including the learner. This incompleteness in both mutual knowledge and self-knowledge is inevitable among humans and may become increasingly relevant for machines also. It also, I believe, accounts for the difficulty of identifying any general panacea for improving human education. A second clear advantage of learning over being programmed is that it offers a limitless potential to recover from errors. A student who seriously misinterprets some concept is likely, at some later time, to discover the inconsistency and recover from it. No omniscient external agent is needed to help.

In this formulation the most important role of a teacher is to point out the next good thing to learn. A second role is that of providing labeled examples. The actual labeling is less fundamental since many natural situations can be considered to be self-labeled. If a cat comes along we may recognize it as a cat from previous knowledge, and hence learn more about cats, without a teacher needing to label it as a cat.

A computational theory of learning also provides the following further wrinkle on the role of a teacher. PAC learning guarantees that a concept class is learnable from random examples for any distribution that the environment may provide. The learner will, however, have a specific learning algorithm. A teacher with knowledge of the learner's algorithm will be in a position to accelerate learning. In particular, the teacher may be able to present a short sequence of well-designed examples that drive the learner's algorithm to the correct concept, after many fewer examples than would be needed if they had been random. For example, if we know that conjunctions are being learned, and the elimination algorithm described in Section 5.5 is being used, then a single stylized example that has only the essential features of the concept, and is devoid of any distracting features, would be effective. This is somewhat like bare bones illustrations in books for very young children. An elephant would be shown with prominent tusks and a trunk, but lacking any other detail. Such illustrations would then have the effect of driving the child's learning algorithm to the correct hypothesis after just the one example.

5.10 Learnable Target Pursuit

The ability of humans to acquire knowledge on a large scale is remarkable. At the basic level this must be based, I believe, on the execution of learning algorithms of the same nature as I have described. Supporting this there needs to be an additional capability that I call learnable target pursuit. At any instant any student, or any entity with a learning ability, has available all those features for which recognition algorithms have been previously acquired. The concepts that can be learned by the learning algorithm in terms of the available features are the accessible targets. They are the targets that will be learned by the learning algorithm, provided that labeled examples of them present themselves. (For this discussion the question of how the recognition algorithms for the currently available features had been previously acquired is not relevant. In biology it would be a combination of evolution and learning. For computers it would be programming and learning.)

I suggest that humans are wired so as to be always ready to pursue any and all accessible targets. This provides a mechanism for continuous learning in any rich environment even in the absence of any teacher. It enables not only previously learned concepts to be fine tuned to higher accuracy, but also entirely new concepts to be learned.

How are examples labeled in the absence of a teacher? As already mentioned, many situations are self-labeled. If I can already recognize both swans and the color black, then I can start learning about black swans if I see one without needing a teacher to identify it. Similarly, if I meet a person I have not seen before, I can start learning about them without a teacher being necessary to identify them or their characteristics.

This capability for learnable target pursuit can operate in the absence of any teacher, but it also provides extra opportunities to be exploited by a teacher. These opportunities include the presentation of examples of concepts for which the student is ready. As already pointed out, progress can also be made without a teacher, but that is dependent more on the serendipity that examples of accessible concepts will somehow appear. Individuals can be smart and seek out experiences that enable them to pursue some useful targets that are accessible given their current knowledge. They can also make the mistake of devoting time to material for which they have insufficient preparation, in which case they may learn little.

5.11 PAC Learning as a Basis of Cognition

I have presented PAC learning as a mathematically rigorous and philosophically satisfactory notion of induction. But I think it is more than that: Its basic features are essential and unavoidable in any attempt to build a theory of cognition.

Humans presumably have some shared learning algorithm. I shall call this algorithm *A*, and the concept class it learns *C*. This observation already provides an account of how we can have shared concepts: Whatever I can learn from examples, I can pass on to you by pointing out examples to you, provided we have the same set of previously acquired features. This may also work if our feature sets are different, as long as the target concept is accessible from both. To date, the algorithm *A* and concept class *C* used by humans have not been identified. But whatever they are, their very existence provides an assumption-free rigorous theory of induction for cognition. It makes no assumptions at all about the world or the distribution of objects in it. The frequencies with which objects have their exponentially many variants with different combinations of properties may be arbitrarily complex and need not be known to the learner. A human having algorithm *A* will manage to learn certain regularities in this complex world, and will miss some others. Other humans will have essentially the same capabilities. Anything you can learn I can learn too, at least in principle.

One much discussed issue in human learning is how some of it seems to occur from positive examples alone. In human behavior it is difficult to establish exactly when this phenomenon can be considered to have occurred. The fact that humans appear to be able to learn to identify members of an animal species from just examples of that species is not conclusive. We may have somehow figured out that each animal belongs to just one species and hence implicitly use positive examples of dogs as negative examples of cats when we are learning the latter. It is therefore not clear to what extent humans do really learn from positive examples alone. But even if we do, there is no inherent mystery in that. As we have seen there do exist learning algorithms, such as the elimination algorithm for conjunctions, that provably do exactly that.

Another issue is that of one-trial learning, or learning from very few examples. As I have previously mentioned, for a fixed algorithm there may be single examples that drive it to the correct hypothesis, and these may be the ones that good teachers provide. There also exist explanations of why a small number of examples may sometimes suffice even without a teacher. In cases where the true concept depends on only a few features among a much larger number of distracting features, so-called attribute-efficient learning is sometimes possible even without a teacher.[17] This means that the number of examples needed for PAC learning is proportional only to the small number of critical features, such as tusks, a trunk, and big ears, and depends only much more weakly on the possibly numerous irrelevant distracting ones. Another possibility is that some concepts may be easy to learn because instances of the category are separable by wide margins from noninstances, in the sense already discussed in the context of perceptrons.[18]

Everything in this chapter requires the Invariance Assumption. In practice, we can never be certain that the world will not change on us in an unexpected way, so that future examples will be from a very different distribution from those in the past. Past performance is not necessarily indicative of future results. Living organisms, however, need to make decisions all the time and take a view on what will happen next. The only course available is to learn as many of the world's regularities as we can, and allow them to guide our decision making. There is simply no alternative.

Chapter Six

The Evolvable

How can complex mechanisms evolve from simpler ones?

> *If it could be demonstrated that any complex organ existed, which could not possibly have been formed by numerous, successive, slight modifications, my theory would absolutely break down. But I can find out no such case.*
> CHARLES DARWIN

6.1 Is There a Gap?

Darwin's *Origins* is a work of breathtaking insight. From observations of plants and animals he deduced a century and a half ago a theory of evolution that has become the central theory of biology. In it, Darwin accepts that the existence of a single organ in biology that could not have been formed by successive small modifications would contradict his theory of evolution. The knowledge gained in the century and a half since confirms his theory in *qualitative* terms. The proposition that current life forms on Earth are related is fully supported by evident similarities among their DNA, as well as by the rich fossil record. The evidence for Darwin's general schema for evolution being essentially correct is convincing to the great majority of biologists. This author has been to enough natural history museums to be convinced himself.[1] All this, however, does not mean that the current theory of evolution is adequately explanatory. At present the theory of evolution can offer no account of the rate at which evolution progresses to develop complex mechanisms or to maintain them in changing environments. It does not explain how evolution on Earth, as suggested by the fossil record, could have occurred within the timescale generally attributed to that record and with the physical resources of the Earth or, for that matter, of this universe. We are faced with a quantitative corollary of Darwin's observation: The existence of an organ that would require too many such modifications to have evolved in the time available would equally contradict his theory.

The theory of evolution through natural selection, as it currently stands, is not of the same nature as some other great theories of science, such as Newton's laws of gravitation or Einstein's general relativity. The latter make quantitative predictions that are subject to verification. In contrast, evolutionary theory at present offers no comparable quantitative predictions, or even quantitative explanations of the past. Perhaps this is why, among the great theories of science, it is the theory of evolution that arouses the most skepticism and organized opposition. Sizable fractions of the population in countries around the world reject it. There is no evidence for any such level of rejection for a spherical Earth, or for the strangest among the successful theories, quantum mechanics. One has to entertain the possibility that evolution's unique position in this regard may be due to the shortcomings of its existing theory. If a quantitative account were available, it seems less likely that significant opposition to it could be maintained.

As we have seen, we do not know in detail how the DNA controls the physiology or behavior of an organism. And if we do not understand how the DNA influences physiology or behavior, we should not expect to understand how changes in the DNA can lead to fitter physiology or behavior. Unfortunately, this gap is not the only problem. As we have discussed earlier in connection with the age of the Earth, Darwin himself was much troubled by the problem of how his proposed evolutionary mechanism could function on the limited available resources. Some others since have also expressed serious concern about the absence of convincing quantitative detail in the theory.[2] Future generations will wonder why these questions have not been asked with greater urgency.

Quantitative theories of population dynamics have existed for a century.[3] These theories are concerned with analyzing competition among static entities, and predicting how their relative population sizes will change as a result of competition. They have relevance to evolution, but do not address the question of how quickly organisms of increasing complexity can evolve.

There is no theory known that explains quantitatively how competition by itself leads to greater functionality and complexity. Yet living things are highly complex mechanisms by any measure. On the part of Darwin, or his successors, of course, there is the excuse that they did not know—indeed, could not have known—what we now know regarding biochemistry and computation. For us, with the bounty of extra knowledge that our predecessors lacked, there are both more opportunities, and more responsibilities, to build a more complete theory of evolution.

Most computational work associated with evolution, namely that involving so-called genetic algorithms, has attempted to invoke ideas suggested by evolution, to obtain better computer methods for tasks such as optimization. This work generally has not been targeted at understanding how evolution works in biology, and in general, these attempts have not produced results that are suggestive of how complex biological systems might have evolved.[4] Darwinian evolution as a panacea for the creation of complex functioning mechanisms remains to be demonstrated. The theory of computational learning, however, rather than simply trying to mimic evolution, can, I argue, help us understand it.

6.2 How Can the Gap Be Filled?

The main question is this: How could complex biological mechanisms have evolved given the time and population sizes that are believed to have been available? Fortunately, much is now known about the nature of biological mechanisms that was not known to Darwin, or to the later formulators of the so-called Modern Synthesis like Ronald Fisher.[5] Human biology we now know is based on more than 20,000 proteins encoded in the human genome. The genome further encodes a network, one that describes the conditions, in terms of the concentrations of proteins present in a cell at one time, that are necessary for a particular other protein to be expressed. It is believed that the regulatory regions in the DNA encode these conditions. Slight faults in the description of either the proteins or the regulatory mechanisms can make an organism nonviable. It is at present quantitatively unexplained how such complex mechanisms with so many interlocking parts can maintain themselves under changing environments and evolve into more complex forms. Of course, Darwin was not in a position to ask this question quite so explicitly. It is quite possible that the reason why no provably effective specific Darwinian mechanism has been identified in the years since his time is simply that insufficient effort has been put into the search.

In order to specify more completely the mechanism responsible for evolution on Earth, one would need to understand how exactly the genomes of each generation are derived from those of the previous one. A simple hypothesis, at least for asexual species, would be that each base pair in the DNA sequence will randomly flip to one of the other three possibilities, with the same fixed small probability. There is no evidence, however, that this simple derivation mechanism is what occurs. Equally importantly, there is no evidence that this simple mechanism can lead to evolution at the pace at which it has occurred in biology. Treating Darwinian evolution as a learning mechanism provides a way forward. This approach enables us to consider not just one possible variation mechanism, such as this one, but many, and to explore the ultimate limitations on all possible such mechanisms. There is every reason to believe that such a more systematic analysis is necessary if we are to understand how evolution can give rise to forms of increasing complexity as fast as it is believed to have on Earth.

Computational learning, as described in the previous chapter, is nothing other than the quantitative study of how computational mechanisms can be acquired without a designer. Life is full of computational mechanisms. If we are to understand how those mechanisms, and life itself, could have arisen without a designer, then computational learning is exactly where we need to look. On the one hand this approach may be disappointing to those who are hoping for biological evolution to derive magical powers from a yet unsuspected source. On the other hand it has the positive aspect that it offers an existing theory on which to build.

Mammalian evolution provides a setting for the kind of understanding this approach might gain us in the foreseeable future. Mammals, which represent more than 200 million years of evolution on Earth, have many proteins that are similar in the different species, and it is likely that many of the important differences among the species result from how regulation of these proteins differs across species, rather than by differences in the proteins themselves. A reasonable first step, then, is to try to understand how species with a fixed set of proteins can maintain themselves and evolve under changing environmental and ecological conditions, under the simplifying assumption that only the protein regulatory mechanisms can change.

In this thought experiment, there are, say, 20,000 variables, $x_1, \ldots, x_{20,000}$, which represent whether or not (or at what concentration) each of the proteins $p_1, \ldots, p_{20,000}$ is present. For each protein p_i there is a so-called input function f_i of the 20,000 variables that specifies whether or not (or at what level) that protein would be expressed.[6] These input functions must belong to some class of the kind we saw in the previous chapters, such as conjunctions or disjunctions.[7]

We can describe these functions mathematically as a set from some class C where, for example, the member f_7 describes the expression level of the seventh protein. How f_7 depends on the 20,000 variables is the question. It may depend on just the three concentration levels x_{23}, x_{136}, and x_{7641}, say. But what is the dependency? It could be

$$f_7(x_1, \ldots, x_{20,000}) = 3.2x_{23} + 0.42x_{136} + 1.03x_{7641},$$

or it could be a similar function with the same three variables but with other parameters. Or it could be a nonlinear function involving quadratic terms, or something completely different. Choosing the class of functions with which to explore these questions presents a stark dilemma, the predicament between Scylla and Charybdis. If the class is too limited, then the biological mechanisms implied will be so restrictive as to be inadequate to express the complex mechanisms needed for life. On the other hand, *if C is too extensive,* then there may not be any Darwinian evolution algorithm that can navigate that complex space of possible functions quickly enough to permit adaptation in the limited time that has been available. In other words, for a class of functions that is too extensive there will be no evolution algorithm, and for one too restrictive there will be no biology. These are the kinds of questions that Darwin and Fisher were not in a position to ask.

The example we use here of input functions for a fixed set of proteins is just a concrete subproblem. The real problem is harder because for that we need to understand other kinds of circuits also, such as those that model evolving proteins. I believe, however, that the more general problem will also be governed by similar impediments, and approachable using the same methodology.

No theory of evolution can be considered complete unless it details the class of functions as well as an actual evolution algorithm that can navigate it. At a very minimum, we need to demonstrate that some plausible candidates for the algorithm and the function class exist. After such a demonstration has been accomplished, the Darwinian theory will have progressed to being more than a metaphor.

Some may question this emphasis on Darwinian mechanisms, citing, for example, Lamarck's theory of evolution, which predates Darwin and entertained the possibility of broad classes of behavior influencing not only the next generation but also later generations. There do indeed exist inheritance mechanisms that function in this way. For example, a pregnant female may influence the fitness of her offspring by eating a poor diet. The sperm and the egg contain more physical material than an abstract DNA string description. At least in principle, there are many avenues for inheritance from parents that depend on information not contained in the DNA.

Such avenues are called epigenetic.[8] Whether epigenetic mechanisms significantly speed up the rate of functional change in evolution is an open question. Fortunately, we can proceed here without needing to resolve it. This is not simply because there appears to be little evidence that these have had a significant positive impact on the long-term rate of evolution on Earth, although that may be justification enough to follow a purely Darwininan model. Rather, this course is justified because any epigenetic inheritance mechanisms that cannot be viewed as Darwinian can also be formulated as learning, though perhaps of a less constrained kind. Therefore, if one hoped to demonstrate that some plausible epigenetic mechanism would yield more powerful evolution than the Darwinian, and would better explain biology on Earth, then computational learning would remain the framework in which to work.[9]

6.3 Does Evolution Have a Target?

I am now getting to the main point about evolution, that it is a form of learning. In order to get there a first immediate obstacle needs to be faced. Most evolutionary theorists deny that evolution has any kind of goal, instead arguing that it is simply a fact that results from competition. Unfortunately, competition itself is insufficient to explain how protein circuits become either more complicated or better fitted to solving objective problems, such as how to see or run. However, learning as we have formulated it *does* have a target function, such as recognizing a species of flower. If evolution is an instance of PAC learning, it, too, must have at least a target, even if not a goal. So what is the target of evolution?

As a metaphor, consider a manufacturing corporation competing with other such corporations. The fact of competition alone does not go far in explaining the corporation's actions. Rather, the insight that the corporation has the target or goal of making a profit, and is not in existence merely for competitive sport, is far more explanatory.

Evolution, as Darwin described it, has a feature analogous to corporate profit: fitness. Fitness in Darwinian evolution is a measure of the benefits that an entity enjoys in some environment. Selection then directs evolution to favor entities with higher fitness. In other words, just as the goal of a corporation is to make a profit, the target of evolution is to maintain or increase fitness.

Fitness has had several definitions over the past 150 years. Many, following the population geneticist J.B.S. Haldane, use it to mean the average contribution to the gene pool of the next generation that is made by an average individual of a specific kind. This notion is appropriate for analyzing the results of competition among static entities, but it does not get at the substance of fitness itself via the factors that influence it. I use the word fitness in the general sense of the phrase "survival of the fittest" as coined by Herbert Spencer in 1864 to describe natural selection, and as adopted by Darwin in later editions of *On the Origin of Species*. One innovation of the approach to be described here is that fitness will be defined in terms of the factors on which it obviously depends, namely the behavior of the evolving entity and the environment. To avoid any confusion with Haldane's description, I call the new measure *performance*, rather than fitness, although it is intended to correspond to Spencer and Darwin's notion of fitness.

Making the assumption already implicit in Darwin's work—that different choices of action have various levels of benefit for the evolving entity—we can define the performance and the target in terms of the notion I call an ideal function. For any species (or other evolving entity), at any instant, in any specific environment, this ideal function will specify in every possible situation the most beneficial course of action. For example, in the protein expression context, the ideal function will specify, for every combination of concentrations of all the proteins, the most beneficial expression level for each protein to go to next.

This ideal function need only exist as an abstract function, and it need not be known to anyone. In the real world this function may be extremely complex. The ideal function in our protein circuit example will refer to the totality of the more than 20,000 functions, each one of which describes the production of a specific protein. (That is, the ideal function would be a set of functions rather than a single function. We shall return to this point later.) As a result, although the evolving entity will have a tendency to evolve toward the ideal function, it will succeed only for ideal functions that come from a simple enough class, just as learning succeeds only for simple enough classes. This formulation entirely finesses our lack of knowledge of how each protein interacts chemically with the cell, as well as what ecological factors bring benefit to the organism in its environment. All this knowledge is summarized in the ideal function.

The target of evolution therefore is simply higher performance. This at once makes the process amenable to treatment as a form of PAC learning. Further, exactly as a machine learning algorithm, or ecorithm in general, the evolution algorithm will succeed without needing any expertise in, for example, ecology. The responses of the learned circuits that realize these behaviors will be theoryless.

The term "ideal" should not be misinterpreted here. There is no implication whatever that the resulting creature is optimal in any sense. There is no suggestion that humans, as we are or could be, or any other extant or possible organism, were the goal of evolution. Ideal is meant only in a very local sense, one evolving entity and one environment at a time. For example, for the human species in the present environment, some behaviors are more beneficial than others. There are better and worse expression functions for our seventh protein, and better or worse amounts of chocolate to eat per day. Since the actions we take in one circumstance may influence what is the most beneficial action in another, it is the combination of all the action functions that is evaluated. The ideal one is that which produces most benefit in that snapshot of an environment.

6.4 Evolvable Target Pursuit

Before continuing with the main issue of how evolution can be formulated as learning, I need to sidetrack to explain the broader perspective within which all this is set. This viewpoint is that of evolvable target pursuit, the exact analog of learnable target pursuit, and it suggests that evolution needs to be thought of as having two components.

First, for any one species (or gene, or whatever we think of as the unit evolving) in its specific current environment there will be a number of accessible target functions that are evolvable by the evolution algorithm in question. If such an accessible target has higher performance than the one realized by the current genome, then evolution will progress inexorably toward that target in a way that requires no unlikely events to occur. This is because the new target function belongs to a simple enough class that its evolution is guaranteed—at least with high probability, as with the accessible targets of general learning algorithms—by an identifiable evolution algorithm that requires only realistic population sizes and realistic numbers of generations. I will leave a technical definition of evolvable, as well as examples of function classes that are provably evolvable, and some others that are provably not, for later.

Second, the questions of whether a specific function, say for the expression level of a protein, has higher performance than another function is determined by complex factors relating to both the current state of the species as well as the current environment. The state of a species can change at any time, if, for example, the input function for a protein changes as a consequence of a mutation in the DNA. The performance of a function can suddenly change also because the environment changes. For example, changes in the relative frequency of the experiences that arise in daily life may change what is best to do, as will also more cataclysmic events in the environment.

Evolvable target pursuit posits that the course of evolution is guided by the succession of opportunities that arise as the species and the environment change. The pursuit of each target is carried out by an explicit evolution algorithm just like the pursuit of a target in learning is carried out by a specific learning algorithm. Hence the course and the speed of an evolution algorithm pursuing its target is predictable in the same sense that this was true for learning algorithms. In contrast, the emergence of a new beneficial target function, through a specific function that was previously deleterious becoming beneficial, has to be regarded as serendipitous, since that may be the consequence of the arrival of a large meteorite or of a volcanic eruption, or something more mundane but still unpredictable.

This two-component view of evolution is consistent with the theory that the rate of evolution of biological forms is variable. The punctuated equilibrium theory asserts that significant changes often occur within short periods, which may be separated by longer periods of relative stasis.[10] Evolvable target pursuit suggests that when a target is accessible and beneficial, convergence toward it will occur at a predictable, perhaps rapid, rate determined by the pace of the evolutionary algorithm. In contrast, changes in the environment that make an accessible target beneficial occur with serendipity at no set pace. For example, convergence to one target may immediately make a second target function accessible, and if that second target is beneficial, then convergence toward that will immediately follow. Or it may be that no second target will become beneficial until a change of climate or the appearance or disappearance of some other species from the environment has occurred.

The idea that the rate of evolution is driven by the rate at which major environmental changes occur is an old one, and was discussed by Alfred Russel Wallace.[11] He raised the question of whether frequent climate changes in the past due to frequent changes in the Earth's orbit might have been responsible for more rapid evolution in earlier times. This question of how frequently targets will become newly accessible or newly beneficial we regard, however, as being outside the scope of evolutionary theory, being determined by extraneous factors. Evolutionary theory may set an upper bound on the rate of evolution, but in an impoverished environment that rate may not be achieved.

The divide between the two components, of convergence to beneficial accessible targets on the one hand, and the appearance of newly beneficial targets, on the other, is made sharp by the apparent hard limits on the classes of functions that are evolvable. These hard limits are analogous to the various limits we have discussed earlier on the classes of functions that are computable, efficiently computable, or learnable.

The evolvable class needs to have some substantial level of combinatorial complexity itself if evolution is to be effective, even if its scope seems restricted when compared with the complexity of the overall structures that occur in nature. If evolvability were possible only for very simple function classes, such as **and** and **or** functions of just two arguments, then we would need many more stages of target pursuit, and with successive targets much more similar to each other, for a complex function to result. The overall requirement is analogous to that found in machine learning, where algorithms are known that can efficiently learn functions, such as linear separators or conjunctions, *over many variables.* In practice, machine learning algorithms for such moderately simple function classes are already effective, because they can automatically discover the relevant variables from among a possibly large number. The challenge that evolution algorithms need to overcome is the same, that of coping with functions of a large number of variables.

In learnable target pursuit the overall course of learning is determined by a succession of small learnable augmentations to what is already known, maintaining the mass of previous knowledge essentially unchanged. This must be the same for evolution, and the evidence for this is direct: Many parts of the genome are preserved essentially unchanged across broad classes of species. Presumably these encode important complex discoveries that are better left untouched, such as the biochemical basis common to life on Earth. Small changes in these encodings would render the organism not viable at all. Other parts of the genome evolve apparently quite rapidly. Preserving rather than discarding well-functioning, useful mechanisms is an essential ingredient of evolution, as it is of learning.

The role of a teacher in a student's pursuit of a learnable target also has parallels in evolution. Students learn from a lecture that has been prepared for their level of knowledge, as target pursuit. Learning an accessible target is a predictable biological phenomenon of their brains that will occur in the heads of different learners in a roughly similar manner if the learners have similar background knowledge. From the student's point of view, what is available to learn is entirely serendipitous. What the teacher chooses to teach is precisely like whatever the environment puts forward. As the prepared student will inexorably (if probabilistically) pursue a learnable target, so will the prepared organism pursue an evolvable one.

This viewpoint also takes us naturally to the issue of modules. In engineering or computer programming it is natural to design complex systems from a number of modules, each of which performs a distinguishable function and interfaces with the others in a clear and simple way. For artificial products modularity is an important design principle, one that offers many clear advantages, such as ease of design, ease of understanding by humans, and ease of maintenance.

Biological systems are also believed to be highly modular. Indeed, that would seem to offer the only chance for us to ever understand them. However, it is not quite as obvious what advantages modularity offers in the case of biology. I suggest that the apparent severe limitations on evolvability may explain it. These limitations allow simple systems, such as those with fewer controlling parameters, to evolve more easily than those that depend on more parameters. Modularity means that a complex system is decomposable into a number of simpler ones that act largely independently. Such simpler subsystems, each evolving separately, would be hindered less by the inherent limitations on evolvability than would a single complex system.

6.5 Evolution Versus Learning

The idea that evolution is a form of learning sounds implausible to many when they first hear it. I will therefore start by recognizing a sense in which the two phenomena are indeed different—albeit in a sense in which evolution is actually weaker than PAC learning. I will go on to argue later that Darwinian evolution can be formulated as a special form of PAC learning.

Consider the following parable. Suppose a species of monkey living in a forest eats any one of bananas, berries, oranges, and apples. This can be viewed as a disjunction x_1 **or** x_2 **or** x_3 **or** x_4, where x_1, x_2, x_3, x_4 represent bananas, berries, oranges, and apples, respectively. Suppose now that a species of berries that tastes bad invades the area. Then a monkey can learn the optimal new disjunction, namely x_1 **or** x_3 **or** x_4, by eliminating the berries variable x_2 from its disjunction after the first experience of a bad-tasting berry. This corresponds to executing the elimination algorithm for learning disjunctions we discussed earlier.

Suppose now, instead, that the new berries are in fact lethal. Can evolution learn to avoid the lethal berries just like learning could avoid the bad-tasting ones? Darwin says not exactly. If a monkey dies from eating a poisonous berry, then from that one event the genomes of its offspring will not be immediately corrected to eliminate x_2 from their disjunctions. The monkey's children do not just reach in and update their own genomes. The mechanism, proposed by Darwin, is that by means of mutations a variety of disjunctions, similar to but different from the parent's x_1 **or** x_2 **or** x_3 **or** x_4, will have been generated also in the various offspring. Some will contain x_1 **or** x_2 **or** x_4 and others x_1 **or** x_3 **or** x_4, for example. It is only individuals with the fortunate DNA—x_1 **or** x_3 **or** x_4—which omits the lethal x_2 from its disjunction, who will survive.

Of course, while this Darwinian mechanism may ultimately achieve the same result of having a population of monkeys in later generations who all avoid berries just as the elimination algorithm would, it is less direct, and less efficient, with more individuals being born who die prematurely. The Darwinian feedback mechanism from the bad experiences to the good genome is, at least, more circuitous than is the direct elimination that is permitted to learning algorithms. Whether the Darwinian mechanism is intolerably inefficient compared to the direct learning method is the basic question that needs to be answered. It is perhaps obvious that in some infinite limit the Darwinian mechanism surely achieves the same result. The real world, as always, is another story.

This Darwinian feedback constraint can be rephrased as follows. In the perceptron algorithm the updates to the hypotheses, as for example in Figure 3.7, can depend critically on the particular examples seen, and even the order in which they were seen. In evolutionary contexts, this would mean that if the last two foods eaten by a monkey were bananas and berries, the order in which they were eaten may influence the resulting genomes of the offspring. That is to say, the perceptron algorithm is Lamarckian. However, in Darwinian evolution, unlike Lamarck's earlier theory, the genetic variants are generated independently of current experiences. The only role of experience is to compare the fitness of the various offspring hypotheses, after they have been generated. In contrast, in general learning a single example can immediately influence all aspects of the subsequent course of the learning process. It turns out that this distinction does make for a demonstrable difference in computational power: There exist concept classes that are PAC learnable, but not when subject additionally to this Darwinian constraint. The resulting substantive question is this: Is Darwinian evolution almost as powerful as PAC learning, or is it substantially weaker in the range of mechanisms that it can learn?

There is a further aspect in which evolution is more onerous than learning. In learning, or at least in machine learning, it is legitimate to insist on being allowed to start the process from any initial hypothesis. We may want to choose the initial hypothesis that is computationally the most effective. In the elimination algorithm for learning conjunctions we always start the same way, with the conjunction of all the variables and their negations. Since the algorithm only eliminates variables on the way, it needs to start with all of them available if it is to have the ability to learn any conjunction.

Evolution cannot afford this luxury. It needs to be able to succeed starting from whatever genome it has. The possibility of reinitializing to a starting point that is algorithmically convenient is not realistic. The problem with moving to such a new starting point is that it may make the species less fit than it currently is by an arbitrarily large amount, and hence not competitive with any cousins who have not made such a move. Biological entities cannot in general afford such arbitrarily large decreases in fitness. Indeed, it is believed that the mutations in biology that have lasted were mostly beneficial or at least close to neutral when adopted. Happily, this requirement of an arbitrary starting point is not totally exotic for learning algorithms in general, and is met by some particular ones. For example, the perceptron algorithm is known to work correctly starting from any initial hypothesis.

6.6 Evolution as a Form of Learning

To see evolution as a form of learning we view the genome in evolution as corresponding to the hypothesis in learning. The performance of the genome corresponds to its expected closeness to ideal behavior, where the expectation is taken over the distribution of experiences the world offers. The goal is to show that if ideal behavior can be represented by a function in an appropriate class, then evolution toward that ideal behavior will occur by means of an evolution algorithm. The course of evolution corresponds to the course of a learning algorithm converging toward a target function.

Returning to the example of the evolution of input functions for a fixed set of proteins, let $x_1, \ldots, x_n$ be the concentrations of the proteins $p_1, \ldots, p_n$, and, for simplicity, let each take only the values +1 or –1, to represent respectively whether they are present or not. For the seventh protein p_7, for example, some function $g_7(x_1, \ldots, x_n)$ will regulate its production. Whether any will be produced will depend on which one of the 2^n combinations of –1, +1 values of the $x_1, \ldots, x_n$ holds.

This kind of function, which takes and returns yes/no values, is called a *Boolean function*. Boolean functions, even for moderate numbers of variables, may be very complex, and only a very small fraction of them, those with short descriptions, can be represented in practice in this world, let alone learned or evolved. For the sake of argument let us suppose that it is the class of disjunctions, of which an instance is

$$g_7(x_1, \ldots, x_n) = x_2 \textbf{ or } x_4 \textbf{ or } x_{11},$$

that are evolving. Some disjunctions in this class may be more beneficial to the owners than others in that environment. The target of a particular evolutionary pursuit would be the disjunction that defines ideal behavior.

The central question then is whether for some useful class of ideal functions, such as disjunctions, there is a resource-efficient Darwinian mechanism that, when started from an arbitrary member of the class will evolve toward the ideal function.[12] To formulate this question more precisely a computational model along the lines of the PAC model with the ideal function as target, and constrained additionally by the limitation of Darwinian feedback, is needed.

6.7 Definition of Evolvability

Discovering the ideal function is like discovering a secret.[13] The evolution mechanism can obtain information about this secret only via the following quantitatively feasible Darwinian process. It will take a polynomial number of genome variants of the current genome and, for each one, take a sample of polynomially many experiences, or inputs to the function. These inputs could specify some aspects of internal chemistry, such as protein concentrations, or external circumstance such as temperature, with the environment dictating their frequency of occurrence. The organism does whatever its genome dictates for the circumstance specified by that input—say, move from shade into sunlight—and the organism will then enjoy the consequent benefit or harm. The average benefit or harm to each genome over the sample of polynomially many experiences is our estimate of the genome's performance.

All this corresponds simply to the owners of the different genome variants going through life, having experiences, and enjoying benefits in relation to how often their life choices correspond to the more beneficial behaviors. The higher the aggregate performance of any one genome variant, the closer it will match the performance of the ideal function, and the more likely that that variant will survive natural selection.

We can define mathematically the performance of a current genome function g with respect to the ideal function f for a given distribution of conditions D. We denote this performance by $\mathrm{Perf}_f(g, D)$. Under each different condition x the ideal action of the evolving entity will be assumed to be either −1 or 1. Mathematically speaking, a genome g has high performance when its output is highly correlated with the ideal function's output over the arbitrary natural distribution D over the set X of all possible conditions or experiences. The mathematical definition of performance then is

$$\mathrm{Perf}_f(g, D) = \Sigma_{x \in X} f(x)g(x)D(x).$$

Here $\Sigma_{x \in X}$ denotes summation over all possible experiences x. The quantity summed is the product of $f(x)$, $g(x)$, and the probability $D(x)$ that experience x actually occurs in reality. (Each $D(x)$ has a numeric value between 0 and 1, and the sum of all those values is equal to 1.) Since g and f take values either −1 or +1, their product $f(x)g(x)$ will have value 1 if f and g agree and value −1 if f and g disagree.

If for every x in X it is the case that f and g agree, then the product $f(x)g(x)$ will always equal 1 and hence $\mathrm{Perf}_f(g, D)$ will equal 1. This is the case when g *is* the ideal function f. In the extreme opposite case, for every x, f and g disagree so that $f(x)g(x) = -1$, and then $\mathrm{Perf}_f(g, D) = -1$. This is the case of the genome doing exactly the opposite of the ideal action in every possible situation. It is easy to see that in all cases $\mathrm{Perf}_f(g, D)$ will take a numerical value between −1 and +1. For example, a value of $\mathrm{Perf}_f(g, D) = 0.9$ is a possibility and would correspond to g well approximating f. The higher this number, the more frequently will the genome be taking the ideal action, where by frequently we mean as determined by the real-world distribution D. The choice of action in the most commonly occurring situations will carry the most weight.

The performance function mathematically formalizes Darwin's notion of fitness. Surely, if the notion of fitness has more than poetical or metaphorical meaning, then we should be able to define how it is related to the factors that determine it. Clearly the actions of an evolving entity as determined by its genome g contribute to the fitness. Clearly also the fitness is determined by the environment. The relevant aspects of the environment are captured both in the distribution D, which characterizes the relative frequencies with which various experiences occur, and also in f, which characterizes the most beneficial behavior of the evolving entity in that environment. Previous theories have not attempted such a relationship. In a later section we shall generalize the performance function to allow behaviors with more than just two outcomes in any situation.

The performance function enables us to more rigorously consider what biologists call the genome fitness landscape. Genomes with higher fitness would be at higher elevations in this analogy. Previous theories have relied on this analogy, and it is a natural setting for discussing how genomes can evolve so as to become more effective, or, in this visual image, reach higher ground. Previous theories, however, have not defined fitness in terms of the factors on which it depends, the genome actions and the environment. The performance function does. To make this theory complete, we will need also to specify how genomes mutate, and how they successfully navigate this landscape in terms of concrete computational processes.

What we have now is a definition of evolution as a form of PAC learning, in which the ideal function f is approached via a sequence of genome functions g, each one of which is selected on the basis of its superior performance among a set of variants generated from the previous genome function in the sequence.

As discussed earlier in Section 6.5, Darwinian evolution does not use the full power available to PAC learning, because of the restrictedness of the feedback. As compared with the PAC model, where hypothesis updates can depend arbitrarily on a single example, evolution selects a new mutation on the basis of its performance, from among several tried. A mutation to a gene may be replicated many times through the descendents of the individual in which the mutation occurred. Individuals with the same gene will compete against individuals with a different gene, and the gene whose aggregate performance averaged over numerous life experiences is superior will eventually outlive the others. Hence, while in a learning algorithm a hypothesis can change because of a single example, in Darwinian evolution it will change—which here means it will be adopted by a population because it is superior—only because of aggregate statistical superiority over many life experiences.

This kind of learning, whereby aggregate measures, rather than individual examples, determine the hypothesis, has been defined by Michael Kearns as the Statistical Query, or SQ, model.[14] In the SQ model, when attempting to learn to distinguish species of flowers, one would ask, for example, approximately what fraction of flowers of species A have petals of length more than 1 unit, rather than ask for the descriptions of individual flowers. These questions may be answered, of course, by taking enough examples of that species and asking what fraction of the sample satisfies the criterion. Hence, PAC learning is at least as powerful as SQ learning. Indeed, it turns out that some PAC learnable classes are not SQ learnable, and hence that SQ is more constrained than PAC.

The evolution model above also asks only aggregate statistical questions, just as the SQ model does, but it cannot ask arbitrary statistical questions, such as about the length of petals. It can only ask questions about one thing—performance, which is the correlation of a given genome function g with the secret, ideal function f. Furthermore, the only way a genome can obtain information about this secret is by having individuals with copies of that genome live. The information about the secret will be revealed only by the survival of those individuals in competition with others.

To summarize, in this model of Darwinian evolution, given a certain genome, a distribution of variants will be generated depending only on the genome and not on any experiences. The selection among these variants will be made by a process that depends on experiences, but only via estimates of their performance $\text{Perf}_f(g, D)$ on natural examples that can be obtained from polynomial-size samples. The model does *not* require that the exact value of the performance $\text{Perf}_f(g, D)$ be accessible. It only assumes approximations to the performance that can be obtained from polynomially many experiences.

As previously discussed, we want the pursuit of an accessible target to succeed from *any* starting genome function g in the allowed class. We cannot expect to be able to "reset" it to something having a form convenient for long-term evolution. The problem with such a fixed reset is that the result may have performance so much lower than that of the current genome that it cannot compete at the start.

Starting from an arbitrary starting point, we want convergence toward f to take place with only a modest-size population and within a modest number of generations, where by modest I mean that they are polynomially, rather than exponentially, bounded in terms of the relevant numerical parameters, such as the number of variables (e.g., proteins). Further, we want the computational cost of the algorithm A that computes the variants from the current genome to be polynomial also. The former reflects the limited time and space available in the universe for the organisms. The latter models the biological mechanism for producing the variants of each generation from the previous one.

Given this model, the main algorithmic design choice lies in how the variants of the next generation are generated. In biology single base substitutions (i.e., changes at a single point in the DNA sequence) certainly occur, but they are not the only source of variation. Whole segments of the DNA sequence are sometimes copied and inserted into another position in the sequence. Indeed, entire chromosomes can be duplicated. Deletions can also occur on a similar scale. In the model here we shall permit all these mechanisms, and much more. We shall allow any polynomial time randomized computation to generate the variants. This may sound overly generous, but as we shall see, even with this allowance, the Darwinian constraint of producing variation independent of experience appears to impose severe constraints on what is evolvable. Computational generosity in the production of the variants is not enough to easily explain away evolution. Note that this generosity is only in the direction of ultimate flexibility in computation, and does not compromise quantitative feasibility. It is defined like this so as to allow for *all* mechanisms that nature might use, even those we have not yet detected, as long as they use only feasible resources. The goal is to uncover *any* algorithm that nature might be using anywhere in the universe. After all, the problem is that no algorithm is currently known that fits the bill. Preconceptions about what is "natural" would be a hindrance, and need to be resisted.

One cannot guess what conclusions this line of work will lead to regarding evolution on Earth. Biology may be using very simple algorithms for producing variants. It may be that certain simple algorithms are more powerful than we currently understand. It is also conceivable that the functions that biology has evolved on Earth are somehow simpler to evolve than they seem. Alternately, the algorithms nature uses to produce the variants are indeed highly sophisticated, and we don't yet have even a glimpse of what they are.

Another way of stating the extreme possibilities is the following. One view of the variation process behind evolution is that it is simply the result of errors occurring in the basic genetic processes. Thus errors in copying the DNA during reproduction could be the main mechanism for producing the variant genomes of the next generation. An opposite viewpoint is that the process of producing the variants is highly complex and clever, as it may well be if it has undergone extensive evolution itself.

6.8 Extent and Limits

It turns out that just as some functions are provably not computable by any Turing machine, some functions are provably not evolvable in the model we have just described. Furthermore, the divide is quite subtle; pairs of similar-looking classes of functions can fall on opposite sides. One such pair, which we will look at in some detail, comprises a special kind of disjunction, called a monotone disjunction, and a class of functions called parity functions.

A monotone disjunction is a disjunction in which no variable can be negated. The disjunction x_1 **or** x_4 **or** x_6, for example, is monotone. It can be specified simply by the set, in this case $\{x_1, x_4, x_6\}$, of variables that occur in it. Given n variables, we can define 2^n monotone disjunctions, as for each variable there is the choice to include it or not. The function's output is "true" or "yes" so long as at least one of the variables is true. For example, a tree might flower if it gets more than four hours of sunlight per day, or receives more than a liter of water per week, or detects a related plant nearby; likewise, a protein might be produced if at least one of three other proteins is already present in a cell.

A parity function is likewise defined with a subset of the variables, such as $\{x_1, x_4, x_6\}$, chosen from the set of size n. Again, there are 2^n possible functions. Unlike the monotone disjunctions, parity functions return "true" or "yes" or "1" if and only if an odd number of the variables in the set are true (or have value 1). (It does not matter whether we call the other possible value 0 or −1 here.)

The problem in either case for an evolution algorithm is to converge to the unknown hidden ideal function, if there is one, from among the 2^n possibilities, with only polynomial resources. The target function in both cases is specified by a hidden subset, for example, $\{x_1, x_4, x_6\}$, that is not known to the algorithm. For both problems there is no chance of testing exhaustively all the 2^n possibilities if n is large, such as 20,000. Assuming that the distribution of inputs is uniform—that all 2^n possible input values occur with the same probability $1/2^n$—we can prove that monotone disjunctions are evolvable. Making the same assumption, we can prove that parity functions are *not*.

A simple algorithm for producing variants can be used to show that monotone disjunctions are evolvable. Variants of an existing disjunction are produced by swapping an existing variable from the disjunction for a new one, deleting a variable, or adding one. These variants are compared as to their performance. It can be shown that convergence toward the ideal function will occur as fast as is required by the definition of evolvability, if the examples are drawn from the uniform distribution. This is not true in general. Vitaly Feldman has shown that there is no evolutionary algorithm for monotone disjunctions (or conjunctions) that works for *all* distributions.[15]

In contrast to this positive result for disjunctions, parity functions are not evolvable in our Darwinian model even for the uniform distribution of examples. To search the set of 2^n possible functions takes exponential effort, and it can be shown that there is no alternative method that is substantially more efficient. As discussed in Section 3.5, proving negative computational results such as $P \neq NP$ is beyond our current capabilities. The evolvability model described here incorporates computation, and yet we have a striking negative result for it. The reason is that the negative result is proved using only statistical or information theoretic arguments. It follows directly from a corresponding negative result of Michael Kearns for the Statistical Query model.[16] Parities are not evolvable, because the statistical constraints on Darwinian evolution, namely that only aggregate behavior on the distribution can be exploited (and not the nature of individual examples), exclude the evolvability of parities. (For comparison we note that, in stark contrast, this same problem is PAC learnable, using an algorithm based on linear algebra that does examine the details of individual examples.)

Of course, any evidence in biology that a parity function on a large subset of variables has been discovered in the course of evolution on Earth would contradict the validity of our model here. However, there seems to be no such evidence, an absence that is consistent and would be predicted by our theory. In fact, any suggestion that the parity function is biologically unnatural only supports the theory.

The fact that we humans have evolved, and can compute the parity of any fixed, known set of 1s and 0s—say, a hundred of them—is no contradiction. For this we need only an ability to count, which is enough to compute whether the number of 1s among these 100 particular numbers is odd or even. Counting is easy. Discovering a subset of variables on which the parity of the number of values that are true is a beneficial criterion of action is much harder.

This claim that parity functions are not evolvable may seem somewhat speculative. Am I really saying that there is *no way* that these things can evolve on Darwinian principles? The answer is yes—there is no polynomial time evolutionary algorithm that will discover an arbitrary hidden parity function with any significant probability of success, if the examples are from the uniform distribution. Of course, when saying evolutionary algorithm here, I have something definite in mind, namely processes that are captured by the definition of evolution I have described. As emphasized earlier, a critical question is whether it is a robust computational model, in the sense that variants of it that attempt to capture the same phenomenon are provably of no greater expressive power. We want some assurance that the results for the model do correspond to properties of a robust phenomenon. Fortunately, there exist some results now that confirm this robustness under variation for Boolean function evolution.[17] Furthermore, even if we consider real-valued functions, as we do in the following section, we still are not able to bypass the impossibility proof.

One can ask whether the assertion that parity functions are not evolvable by any Darwinian process is of a similar nature to Turing's assertion that the Halting Problem is not computable. Both statements are proven mathematical facts about certain computational models. As we have mentioned earlier, Gödel regarded proven statements about the computable as "absolute" statements in the sense that they are independent of the formalism used. Will our assertion about the nonevolvability of parities ever attain a similarly absolute status? The answer to this hangs on the question, I believe, of whether the models of evolvability for which this negative statement can be proved will be ever accepted as robust models of Darwinian evolution, as has been the Turing machine for computation. We shall see in the coming section that robustness for evolution is not quite so simple.

6.9 Real-Valued Evolution

One limitation of the formulation so far given is that it refers only to functions that are Boolean, or have just two possible values, whether true/false, yes/no, or 1/0. Whether protein networks are Boolean rather than real valued is not yet well understood, but a tentative answer would be yes and no. For some proteins the message may be simply whether or not they are being expressed, but for others the amount being expressed seems important. This latter case provides strong motivation for extending the evolutionary model to allow for numerical or real-number quantities, rather than just Boolean true and false.

There is also a second reason, which springs from the fact that there has been little success in finding Boolean classes that are evolvable for all distributions, and not just for particular ones such as the uniform distribution. As we have seen, even conjunctions are not so evolvable. The effectiveness of machine learning in practice depends heavily on the use of algorithms that are resilient to variations in distributions, and are not overly specialized to one. Learning would not be a powerful natural phenomenon if each real-world distribution would require a separate algorithm. This resilience to different distributions is surely equally essential for any plausible account of evolution.

Going beyond Booleans, however, introduces some new complications. If your protein is being expressed at level 3.7, say, it is one thing for the environment to penalize you a fixed amount according to whether 3.7 is suboptimal. It is quite another for the penalty to depend on how far the 3.7 is from the optimum, which may be 1.8, for example. A world in which the penalties can take on general numerical values in this way may be able to provide more detailed and nuanced feedback to the evolving entity. It may therefore support more powerful evolutionary mechanisms. However, for such a more general model the performance function needs to be redefined.

That is an interesting task. For each situation x the penalty or loss for having the genome give a value $g(x)$ rather than the ideal value $f(x)$ will be $\text{Loss}(g(x), f(x))$, where Loss is some function. Unfortunately, that Loss function could take on many forms. Perhaps it is the difference of the values—in the above example, 1.9. Perhaps it should be the square of the difference, namely 3.61, or perhaps something else again. If f and g can take only two values (such as −1 and 1), then the difference is either 0 or a fixed nonzero value (2 in this case). With real-valued functions the question of what loss is suffered by not taking the optimal action is determined on a case-by-case basis by the environment and the evolving entity. The loss function may vary for different genes, and it is not clear what loss functions are reasonable. It is in the essence of PAC learning theory to make minimal assumptions about the world. To obtain such a minimal-assumption theory, it would seem that we would need to show that evolutionary algorithms are robust in an additional new sense also—of being convergent for very general classes of loss functions. As we shall see soon, this is indeed possible.

Suppose now that we have some loss function. How are we to define performance? We shall again take the sum of the losses for the different values of x (e.g., protein concentration combinations), each weighted by the probability of that combination occurring. This can be stated equivalently as a probabilistic expected value of the loss if the combination x is chosen randomly from the distribution D of all possible events:

$$\mathrm{Perf}_f(g,D) = \mathrm{ExpectedValue}_{\underline{x} \in D}(\mathrm{Loss}(g(x), f(x))).$$

Note that in this real-valued definition of the performance function, with the loss function taking the place of the correlation function, performance goes down as the g gets closer to f, while it goes up in the first definition. This difference is for the sake of convenience.

The first steps in investigating the power of real-valued feedback were taken by Loizos Michael in the context of learning Boolean functions.[18] His intermediate hypotheses took numerical values, but the goal at the end was to compute an approximation to the target Boolean function. This approach was further developed extensively by Vitaly Feldman, who showed that for a wide class of loss functions all of SQ could be evolved for all distributions.[19] This is a strong result regarding the power of evolution since it shows that evolution is comparable to a significant learning class. It assumes, however, that the organism evaluates and gets feedback on the function class of its real-valued hypotheses, which may be very different from the final Boolean target class.

Paul Valiant has given a definition for a model that is entirely real valued and that has the important further feature that it applies to a *set* of genomes $\boldsymbol{g} = g_1, \ldots, g_n$ and a set $\boldsymbol{f} = f_1, \ldots, f_n$ of ideal functions.[20] This is needed if we are to model many functions simultaneously, such as the 20,000 or so expression-level functions of our proteins. After all, it does not make sense to talk about the ideal function of a single protein in isolation, since as expression levels of the other proteins change, the ideal behavior for the first protein may change as well. To take a more everyday example, if one wants to reduce one's calorie intake to a certain level, then reducing the number of meals and reducing the amount of food per meal are both relevant actions, but neither can be optimized without knowledge of the other.

For both the protein and the eating examples, it is meaningful to discuss the ideal function set ***f*** for the set of all the functions in their entirety. Furthermore, in the real-valued setting it is natural to discuss the loss function in this multidimensional setting, as illustrated in Figure 6.1. Here the values of the function sets ***f*** and ***g*** will be points in the high- (perhaps 20,000-) dimensional space determined by the values of all their constituent functions. The loss for an entity in acting according to the values determined by its genome rather than the ideal values will then depend on some measure of how far apart these two points are. As when dealing with individual genes or functions, this loss may be the distance between the two points, the square of the distance, or some other measure.

Within this model Paul Valiant has shown some positive results of considerable generality.[21] In particular he showed that natural classes of real-valued functions, including linear functions, such as $f = a_1x_1 + \ldots + a_nx_n$ (and more generally, multivariable polynomials of constant degree), are provably evolvable for all distributions in which examples occur in a bounded region, and for a very general class of loss functions, namely all convex loss functions. (Both the distance and the square of the distance functions are examples of convex functions.) Further, these results apply not only to single

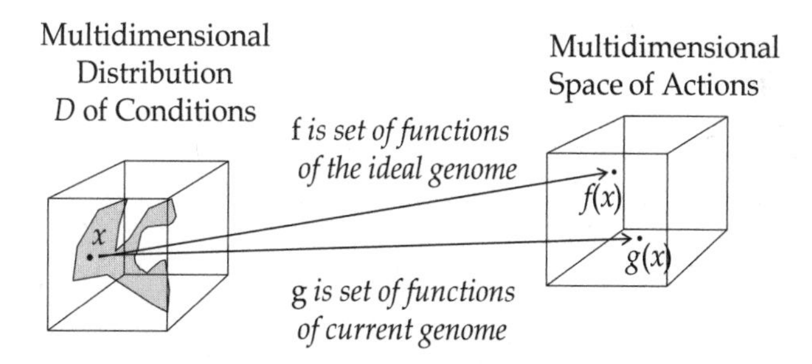

Figure 6.1 The diagram illustrates the definition of evolvability when both the variables and target function take real number values. Besides being more general than the Boolean model, a conceptual advantage is that it becomes natural to consider a set of functions acting simultaneously on an input *x* and to evaluate the loss that that combination suffers as compared with the ideal combination of actions. For example, the set of input functions specifying the expression levels of some 20,000 proteins can be evaluated for their combined fitness. For the combination one can quantify the loss as some measure of the distance between the ideal combination *f*(*x*) and the actual one *g*(*x*) in this high-dimensional space.

functions, but to sets of functions evolving simultaneously, say one for each gene, in the sense we have described. The existence of positive evolvability results of this generality lends credence to Darwinian evolution being indeed viable in the concrete computational formulation given for it here.

Many questions remain unresolved, in particular the following tension. On the one hand some evolutionary algorithms are simple and efficient but only limited generality can be proved for them, as far as distributions or loss functions, for example. On the other hand there are algorithms that are less efficient, having larger polynomial bounds, but are provably effective in greater generality. Some very simple algorithms can be proved to be effective for the particular case of the quadratic loss function, while only more intricate algorithms are known for more general cases. A better understanding of the power of evolutionary mechanisms remains an important goal for the future.

6.10 Why Is This Theory So Different?

In his *On the Origin of Species*, Darwin sought to marshal all the arguments and evidence that could be found in support of his theory of evolution. The titles of his chapters describe the material he used: "Variation under Domestication"; "Variation under Nature"; "Struggle for Existence"; "Natural Selection"; "Laws of Variation"; "Instinct"; "Hybridism"; "On the Imperfection of the Geological Record"; "On the Geological Succession of Organic Beings"; "Geographical Distribution"; "Mutual Affinities of Organic Beings: Morphology: Embryology: Rudimentary Organs." He also has a chapter called "Difficulties on Theory," which discusses issues such as speciation: Why do organisms cluster into distinct species, rather than there being a continuum of varieties? In later work he studied the role of sexual selection. He used all these angles to argue that natural selection is consistent with all the evidence, and is the simplest explanation of the evidence.

But remember Paley's objection, cited in Chapter 2, that life forms are just too complex to have evolved. Darwin avoided confronting this issue directly, as have also his successors. It is one thing to demonstrate that natural selection is qualitatively consistent with the evidence; it is quite another to show that some concrete realization of it is consistent with the evidence in quantitative terms.

A weakness of the current conventional wisdom in evolutionary theory is that it gives centrality to competition per se without proving that that mechanism is sufficient. Competition may be essential, but saying that does not explain everything. One needs a theory that explains why competition increases functionality. We need to explain how evolution is possible at all, how we got from no life, or from very simple life, to life as complex as we find it on Earth today. This is the BIG question. One also needs a theory for the related question of accounting for the rate at which a circuit can adapt to changing environments. Our formulation of evolution addresses the latter question directly by giving bounds on the resources needed to evolve from one function to a better one. It also partially addresses the BIG question by giving bounds on the evolution of a new added circuit. Clearly various different circuit types play a role in biology, and it is not clear how all of these should be formulated for evolution. For example, we do not know how the evolution of new proteins should be formulated in circuit terms.

There are many differences between the model of evolution described here and previously studied models. Most previous models emphasize that evolution takes place in populations containing a broad diversity of genomes. Here I have one genome that generates a number of variants, and after some competition, one genome emerges as the sole winner, and then the cycle repeats.

It is reasonable to ask why diverse populations are not mentioned in this model. The answer is that it can be added, and should be if some extra power for it can be demonstrated. If one could prove that such population effects give a richer class of functions that are evolvable, then that would be a significant result. At the moment the indications are that, while this is not the case in general, diversity in populations with sexual reproduction may allow the number of generations to be reduced at the expense of a larger population size.[22]

The one-genome model may nevertheless be sufficient. We can simply interpret the one genome not as the genome of one organism, but as the sum total of all the genomes of the many individual organisms in a population. The advantage of this is that there are many different ways that the genomes of individuals in a population can interact, and these can all be described as appropriate mechanisms within such a population. For example, both asexual and sexual reproduction have been successful on Earth, even though the latter is considered to be the more successful.[23] A third form of interaction is lateral or horizontal gene transfer, where information is exchanged between individuals who are not descendents of each other. Transfer of genes between different bacteria is believed to be common, and to be a factor in the spread of resistance to drugs. Some consider horizontal gene transfer to have been important in the evolution of single-celled organisms. All these forms of gene transfer can be regarded as computations on the sum total of the genomes of a population, as can also the mutations of individual genes. The one-genome model can embrace all of these by having different internal mechanisms for producing variation.

There are many aspects of evolution that we do not address at all. Diversity in the gene pool may be a good defense mechanism against the unexpected and hence be critically important for survival. Survival is no doubt indispensable, but by itself it does not explain increasing complexity—not all mechanisms that are needed for survival are necessarily useful for enabling increasing complexity. The analysis here can be viewed as one that isolates the imperatives of complexity in evolution from the many other facets of biology.

Another aspect of our theory is that it presupposes a static world. It considers how a phase of target pursuit can be accomplished while the world is kept fixed. This is consistent with the notion that once a target becomes accessible and beneficial, evolution toward it will proceed quite predictably and rapidly. However, the theory can be adapted also to slowly changing worlds.[24]

I have sought a solution from the study of learning. This in retrospect is an obvious place to look. After all, machine learning is the general field that studies how complex mechanisms can be created without a designer. Darwin and Wallace were investigating a very important but special case of this.

Darwinian theory now pervades biology as well as many other disciplines. In biology evolution is identified not just with the Darwinian mechanism, but also with its apparent by-product, the history of life on Earth. This history has been filled with much drama, from the Cambrian Explosion and the Permian Extinction, to the appearance of creatures that can launch themselves into orbit around the planet. My concern here has not been with chronicling the history of these events. It has been only to understand one question: How can any mechanism account for this remarkable unfolding drama?

Chapter Seven

The Deducible

How can one reason with imprecise concepts?

True genius resides in the capacity for evaluation of uncertain, hazardous, and conflicting information.
WINSTON CHURCHILL

7.1 Reasoning

The tension between reasoning and learning has a long history, reaching back at least as far as Aristotle, who, as already mentioned, contrasted the "syllogistic and inductive" in his *Posterior Analytics.* In his treatise, however, Aristotle dealt almost entirely with the syllogistic, which may have triggered the high regard Western civilization has had for reason ever since. More recently, logical reasoning has fallen from the pedestal of highest repute. As has been pointed out often, humans are bad at logic. But that is not the only problem. Computers are very good at logic, yet we do not typically trust them for evaluating uncertain, hazardous, and conflicting information—and even when we do, the computer systems that succeed in this are usually based not on logical reasoning but on learning from large amounts of data.[1]

This chapter adopts Aristotle's dictum that beliefs come from two fundamental sources: syllogism and induction, or reasoning and learning. Despite the beating that logic has taken in recent years, my goal is to describe how these two sources can be unified into a consistent whole. In doing this, primacy will be given to learning, but reasoning will still remain essential.

In previous chapters, I have distinguished subject matter that is theoryful (in the sense that an explanatory theory of it is known) from that which is theoryless, and I have argued that beliefs about the theoryless have the semantics of PAC learning because they are acquired, in the first instance, inductively by learning. (Once one individual has learned such a belief, it may be transferred to another, but even then the semantics remains that of learning.) I shall now address the question of how reasoning with such theoryless information can be justified at all. This is of some relevance since, more often than not, humans reason about theoryless subject matter.

That reasoning makes sense and is theoryful for theoryful content has been long established by mathematical logicians. Following the work of George Boole and Gottlob Frege in the nineteenth century, much progress was made in understanding the forms that a mathematically rigorous view of reasoning can take. We can now ask questions about the nature and power of reasoning mechanisms within mathematics itself. Notions of truth and provability have been defined and distinguished, and questions have been asked as to whether, in a given system of logic, everything true is provable, and everything provable is true.

The mathematical logic that developed from this work has proven to be applicable to formal subject matter, particularly mathematics itself, and to have something significant to say about such subjects. By the beginning of the twentieth century the concerns of this field had moved center stage in the intellectual arena. Can all mathematical questions be translated into one unified formal language? Can any true mathematical statement so expressed, but no false ones, be deduced from a common set of axioms whose truth is self-evident? Some notable figures, including the philosopher Bertrand Russell and the mathematician David Hilbert, had believed that such a program could be carried through. To widespread astonishment, in 1930, twenty-four-year-old Kurt Gödel showed otherwise. In particular, he proved that in any rich enough logical system there were true statements that were not provable. This development, negative as it may have seemed, had profound intellectual impact. It was perhaps the first result that gave a glimpse of some ultimate limitations of what could be achieved by reasoning, even in a completely theoryful arena. Perhaps most importantly it led within a few years to the investigations of Turing and others into computation and *its* limitations.

These discoveries in mathematical logic, however significant they may have been in their own right, did not address directly the problem of reasoning about the theoryless. It was left to researchers in artificial intelligence, from the 1950s onward, to attack that problem.

The logical approach to artificial intelligence, pioneered by John McCarthy, treated the theoryless essentially as if it were theoryful. Axioms were to be constructed for any concept for which a word could be found in the dictionary. This included everyday concepts that were well outside the domain of any known science, and for which such axiomatization had never been attempted before. Rules of inference were then applied that were sound in the sense that they never yielded false conclusions when used within appropriate formal systems. This approach and its equivalents became the conceptual basis for much of the early work in the newly established field of artificial intelligence. More recently, analogous approaches have been pursued in probabilistic formalisms, where the rules of inference are those that apply in probability theory. These approaches treat the subject matter also as being theoryful, making them broadly equivalent to the logical approach.

While both logical and probabilistic modeling are mathematically principled when applied to the theoryful, they offer no principled guarantees when it is not clear how the models relate to the underlying reality, which is the case when the subject matter is theoryless. From a learning viewpoint, however, as we shall show, one can salvage some guarantees on the results of reasoning, even in this unpromising setting. The guarantees that can be achieved through learning are in the qualified PAC sense that, while errors are inevitable, their level can be controlled by putting in an effort commensurate with the quality of the guarantees that one is seeking.

It is important, I believe, to make a clear distinction between the two approaches—mathematical modeling using some kind of logic or probabilistic model, as opposed to learning. In practice it is easy to blur the difference by mixing the two. For example, consider a model, intended for use in a speech recognition system, of how people pronounce the words "yes" and "no." Such a model will be typically probabilistic. The question is whether the model is to be entirely programmed. If it is, then this would be treating the phenomenon as theoryful since one is attempting to have a model or theory of the outside reality. However, more often than not, after such a model has been designed its numerical parameters are tuned by learning. If the resulting model turns out to be useful, the question arises as to whether that success was due to the learning or to the initial programmed model. As another example, we may start with a model for medical diagnosis that first incorporates some beliefs derived from interviews with physicians about the relationships among various physical conditions and symptoms. When tested against real data, the parameters that represent these relationships may have to be revised.

Regardless of the starting point, once we embark on learning from data we have to acknowledge that we are seeking the benefits of the learning phenomenon. It is then most appropriate to evaluate success by the criteria of accurate and efficient learning as described at length in Chapter 5. What is less clear, if success is to be measured by the criteria of learning, is why we need reasoning at all. There may be a portfolio of tasks, such as recognizing dangerous animals, walking up steps, or uttering appropriate greetings, that is sufficient for life. And perhaps these tasks can all be learned.

7.2 The Need for Reasoning Even with the Theoryless

For simple creatures it may well be that a repertoire of learned responses is sufficient. The application of a single learned circuit, which I shall call a reflex response, approximates the best behavior for a specific situation in life.[2] (Here, the word *circuit* is used in the same general sense in which I introduced it in Chapter 4.) Certainly, even for humans, such a repertoire of reflexes is often sufficient. In driving a car, experienced drivers are believed to cope by invoking reflex responses learned from similar previously seen situations. They do not need to go through an explicit reasoning process that considers the possible alternative sequences of events that would follow from alternative actions. Nevertheless, these reflex responses are not sufficient to explain all of intelligent behavior. There is a place for something more, and that something is reasoning. If, unlike mathematical logic or probabilistic reasoning, this reasoning is to be compatible with theoryless knowledge, it will need to be able to manipulate uncertain and unreliable knowledge in a principled way, so that some guarantees are provided on the accuracy of predictions.

The simplest form of reasoning that meets these demands is the application of two learned circuits, each having the semantics of PAC learning, in succession. Chaining two or more circuits may be effective in a range of situations in which it is unreasonable to believe that a single circuit might have been learned.

Consider the following example. If you are asked whether Aristotle ever climbed a tree, or whether he had a cell phone, you probably can answer both these questions with some confidence, despite not having previously had an opportunity to gain much statistical evidence for either question directly. It is implausible that in answering either of these questions you are invoking a single circuit that has been learned to recognize instances of some single concept such as cell phone ownership or tree climbing. Having a reflexive response to such a question would require a specially trained circuit in your brain that takes a name as input and outputs whether that person owns a cell phone or ever climbed trees. Implausible, indeed. It seems more plausible that you rather apply a sequence of circuits, each encapsulating a common sense rule. In this way you successively add more and more information to a picture in your mind. You would start by identifying Aristotle as a particular human who lived in a certain era. You would then apply some common sense rules in succession to make some deductions in order to build up a more complete picture. Ideas about children liking to climb trees and the era in which cell phones were invented would be expressed in these rules. Some of these rules may express facts, while many have theoryless content and will have been learned inductively.

A possible criticism of this example is that humans do not learn any such rule about tree climbing and cell-phone ownership simply because the issue is not of vital value to us. If making such decisions instantly were important, then perhaps we would learn reflex responses for them. Hence the possibility remains that reasoning is only useful for contrived questions of little importance. If animals survive in the main by performing tasks that are all reflex acts, then perhaps human reason is useful only for arcane puzzle solving and is not so fundamental after all.

However, more basic arguments can be put in favor of the extra power of reasoning. The chaining of learned circuits may enable capabilities unattainable by the application of a single learned circuit. For example, the needed single circuit may be of a form that falls outside the class that is learnable, but it may be equivalent to a pair of circuits that each can be learned separately. So, to continue with Aristotle, once we have recognized him as being human, there may be a simple condition to decide whether he ever climbed trees. However, without first identifying the species of the individual in question, a general criterion for what can climb trees may be too complex to be within the learnable class.

As with PAC learning itself, the conclusions derived through reasoning will be permitted to be wrong sometimes, simply because the learned rules are permitted to be wrong sometimes. We can never be absolutely certain of our conclusions; Aristotle may just have aroused enough curiosity in the cosmos to have attracted extraterrestrial visitors bearing cell phones. Nevertheless, even in the presence of all these uncertainties, principled reasoning with some guarantees of accuracy is still possible. Furthermore, both the strengths and weaknesses of the guarantees are important.

7.3 The Challenge of Complexity

Several challenges to the endeavor of understanding reasoning have been encountered in the course of a half-century of research in artificial intelligence. I regard four—computational complexity, brittleness, semantics, and grounding—as the most pertinent to the approach presented here. My approach to addressing all these challenges is that of ensuring that there is some unambiguous relationship between the information represented in the reasoning system and what this representation refers to outside of itself. The relationship for this will be the same as for PAC learning, and will be called PAC semantics.

The first challenge is the impediment of noncomputability and computational complexity. In Section 5.1 I quoted Turing's reference to this, in the context when reasoning is equated with mathematical logic. Similar issues of complexity have been found to arise in other formulations as well. Indeed, Turing's proof that the Halting Problem is not computable can be viewed as an early warning of what has been called the doom of formalism. Expressing what we wish for in a formal framework is often futile if that framework is too broad to permit efficient computation. I do not accept that formalism itself is doomed. The challenge is to identify a formalism that works—one that is extensive enough for the task at hand, without being so extensive as to be computationally intractable. Just as for learning and evolution, we have to sail again between Scylla and Charybdis. Given the extent of our discussion of the issue of computational complexity already, I will leave this without further discussion here.

7.4 The Challenge of Brittleness

Over the decades extensive efforts have been made to imbue computers with knowledge so as to enable them to answer common sense questions. In the main the knowledge has been programmed by humans, and a formal logical system has been used to reason about this knowledge.[3] Overall, these efforts have had only limited success to date, and in situations not foreseen by the programmer computers regularly fail, giving responses that have often been unreasonable or even absurd. The impediment of computational complexity by itself does not account for this failure. Even when the intended, mathematically principled deductive processes have reached a conclusion in the time allowed, the computer's conclusions often fall short.

The reason for such failures must be that the programmed statements, as interpreted by the reasoning system, do not capture the targeted reality. Though each programmed statement may seem reasonable to the programmer, the result of combining these statements in ways not planned for by the programmer may be unreasonable.

This failure is often called *brittleness.* Regardless of whether a logical or probabilistic reasoning system is implemented, brittleness is inevitable in any system for the theoryless that is programmed. If the represented information is not consistent within itself and in exact correspondence with the domain being modeled, then no claims can be made about the accuracy of the deductions. While these systems are mathematically principled for theoryful content, they offer no useful guarantees for the theoryless.

The only way of avoiding this brittleness and achieving robustness is to have the systems learn. The predicate calculus and Bayesian probability are both well-founded mathematical systems. However, within these systems the issue of robustness in controlling errors in the face of limited data and limited computation is not addressed. Learning theory does address these very issues and is therefore a more appropriate basis for this enterprise.[4] Indeed, the advantages are two-fold, at least. First, PAC learning, by definition, is concerned with robustness to computation and data. It quantifies how the accuracy guarantees of the learned rules get stronger and stronger with more and more data and computation. Without some such guarantees little can be said about any system that is less than perfect. Second, a learning system has the fundamental advantage that it can check its predictions against the world. If it finds that it is making false predictions, it can adapt itself so that it will be more likely to be accurate in the future. With such feedback the system can recover from almost any gaps or inconsistencies in its knowledge.

A closely related issue is that of resilience to noise. Humans can cope even in situations when some of the information provided is false, perhaps corrupted by noise during transmission. Happily, the basic PAC model, which does not discuss noise, is easily extended to accommodate it. Further, it has been shown that learning algorithms can be made resilient to certain kinds of noise in some generality.[5] For example, if one introduces noise by randomly changing the label of the examples with some probability, then classes that are PAC learnable in the absence of noise often (and SQ learnable classes always) remain PAC learnable in its presence.

Empirical efforts toward endowing machines with common sense knowledge have shown that the amount of knowledge needed is much higher than was ever expected. This is a yet further source of difficulty, but its sheer scale points even more forcefully to the need for principled automation, and hence learning, in the knowledge-acquisition process.

I do not claim that the learning-based approach I advocate here is without its challenges. Studies of learning and reasoning have shown that unless these problems are formulated very carefully, the computational complexity of each may become too large to be tractable. I shall come to a suggested resolution to this question, but not before discussing the remaining two challenges.

7.5 The Challenge of Semantics

I believe that no system that reasons with large-scale general knowledge can effectively work without there being a clear correspondence between the information represented in the system and the outside reality to which it refers. To understand or construct such systems one needs a principled view of this correspondence. It seems unreasonably optimistic to adopt an unprincipled view, and expect to meet this most basic requirement by pure chance.

If one programs a machine in terms of everyday concepts expressed in English, then one needs to be sure that each word is used in a consistent way throughout. Almost any word in a natural language has some range of meanings, and some words have several distinct meanings. If in a programmed rule words such as port, green, or conservative are used, then one has to be sure that every rule uses these concepts in exactly the same sense.

If several meanings of "port" are to be distinguished—a drink versus a mooring place, for example—then the variants have to be named (e.g., port1, port2) and used consistently. The difficulty of doing this consistently enough accounts for a significant aspect of brittleness.

PAC learning offers an approach to addressing this problem. At each instant for each concept, such as port2, the system will have a hypothesis or program for it that recognizes it, in the sense of saying, "Yes, this is an example of port2," or "No, it is not." This program will have been learned in terms of features that were already recognized. Some of these features may have been themselves learned previously. Others will have been preprogrammed, or, as is the case for living organisms, learned through evolution. Examples of this latter category are light detectors in the retina or pixels in a camera input device for a computer. But, whatever these features are, the end product in the overall system will be a recognizer for the concept of "port2." Inputs presented to the system can be labeled by this recognizer to indicate which are examples of "port2" and which are not.

If all the recognizers in a system are largely consistent in the sense that in most natural situations the labels attached to the inputs do not contradict what the recognizers say, then we can consider the system to be consistent in the PAC sense. If, however, contradictions are often encountered in natural situations, then the system can detect this for itself and seek to reach a more consistent state by modifying its recognizers. An important goal of learning is to reach PAC consistency in this sense.

For such PAC consistent systems the meaning of a concept is simply whatever the circuit labeled by that concept recognizes. Thus, after training to PAC consistency, the meaning of the concept port2 in such a system is nothing other than the function computed by the circuit that has port2 as its target. This circuit may involve other learned concepts but ultimately will depend on preprogrammed features that take external sensory inputs. The relationship between the function this circuit computes and the outside reality is one of PAC semantics.

7.6 The Challenge of Grounding

Finally we arrive at the fourth challenge, which I call grounding. It is intimately related to both semantics and brittleness, and it deals, to put it simply, with two primary issues: the scope of the knowledge that is claimed to be represented, and the constraints of time, space, or other limitations within which the PAC semantics are to be accurate.

These concerns arise even for the simplest of logical statements. Consider the assertion, "All humans are mortal." A logician might phrase it, "For all *t*, if *t* is human then *t* is mortal," and abbreviate it as

$$\forall t \text{ human}(t) \rightarrow \text{mortal}(t),$$

where the inverted A, the ∀ operator, denotes "for all." Even with this simple example there are some obvious difficulties. What exactly is the range of *t* that this statement applies to? If it is to apply everywhere in the universe, how can we be so sure that the assertion is true? Does it apply to humans described in fiction? This last question is not irrelevant if we wish, for example, to learn about the world from written text, since much text refers to fictitious individuals. The fact that Superman is fictitious is good to know if we want to learn only about real people.

We can always add some preconditions to statements that specify in more detail what scope is being asserted. For example,

$$\forall t \text{ nonfictitious}(t) \rightarrow (\text{human}(t) \rightarrow \text{mortal}(t)).$$

This would ensure that human mortality is asserted only for nonfictitious humans. However, we have an infinite regress here. Are we sure this is a complete definition? More to the point, how do we define the terms nonfictitious, human, and mortal? Will not these have the same problem? Note that the complementary existential quantifier ∃, the mirror image of "E," which denotes "there exists," raises all the same issues.

The severity of this challenge can be appreciated even more if we recognize that human intelligence is applied effectively every day to issues with much less universality and permanence than this example. We interact with different people with different personalities and desires. We need to predict their behavior without having axioms that describe what they will do and under what circumstances.

We need a principled basis from which to approach this problem of grounding. PAC learning addresses this by identifying a specific distribution *D* with respect to which it learns and performs. Without some such notion of grounding there cannot be a theory of learning. But how are we to specify this distribution D of typical situations from which a human individual learns? How is it ensured that in frequently occurring situations we will do the right thing, and how are we protected from making rash decisions that are not supported by our experience? Whatever the mechanism is, we need the same kind of protection in artificial systems. If we create a system to perform a task, but have no target distribution in mind on which we expect it to behave well, then we really have no idea what we are trying to accomplish.

To press the point in a different way, we conclude here with a paradox that illustrates the fallacy of discussing probabilistic events without defining what distributions they refer to. Consider the following proposal, and decide whether it would be profitable for you. You are to go up to a random person in the street and offer to compare the amount of cash you each have on you and agree that whoever has more gives it to the other. You convince yourself by the following (false) argument that this will be profitable. You argue as follows: "I have x amount of money. The other either has more, say y, or less, say z. (Ignore the case that they are equal, when there is no gain or loss.) The two possibilities occur with the same probability 0.5. If the other has more then I win y, otherwise I lose my x. Hence my expected gain is $0.5y - 0.5x$, which is greater than zero since y is greater than x." Would you get rich by repeating this? Since the other person could argue the same as you, and it is not possible for both players to have an expected win, the argument just given must be fallacious. But which one is the fallacious step in the above argument? Common sense gives a clue. In practice you would not play the game if you had just taken money out of the bank, but might if you were on your way to do that.

7.7 The Mind's Eye: A Pinhole to the World

In this section and the next I shall describe robust logic.[6] It is an approach to the reasoning problem that addresses all four of the obstacles. It formulates learning and reasoning with a common semantics, maintains computational feasibility, and provides a principled approach to the problems of brittleness and grounding. The device that enables all these issues to be addressed simultaneously is a quantitative formulation of working memory, a notion originally proposed in less computational terms by cognitive scientists. I shall call this computational version the mind's eye.

The notion that in the process of thinking we employ some special mechanism, other than our general long-term memory, for bringing together the different threads of the subject we are thinking about is a central idea in cognitive science. The mechanism is called working memory, and it is closely related to other notions such as short-term memory, imagery, attention, and consciousness. It has been researched from numerous perspectives, and one of the most striking findings is how limited it is. Its restrictedness was memorably demonstrated by the cognitive psychologist George Miller, who in his celebrated paper "The Magical Number Seven Plus or Minus Two" showed that we could hold only about seven objects simultaneously in this working memory.[7] The mind's eye, as colloquially used, is this same notion. When we are thinking, we are usually aware of very few things at a time. For our discussions it will suffice to recall the following earlier piece of introspection offered by the nineteenth-century polymath Francis Galton:

> When I am engaged in thinking anything out, the process of doing so appears to me to be this: The ideas that lie at any moment within my full consciousness seem to attract of their own accord the most appropriate out of a number of other ideas that are lying close at hand, but imperfectly within the range of my consciousness. There seems to be a presence-chamber in my mind where full consciousness holds court, and where two or three ideas are at the same time in audience, and an ante-chamber full of more or less allied ideas, which is situated just beyond the full ken of consciousness. Out of this ante-chamber the ideas most allied to those in the presence-chamber appear to be summoned in a mechanically logical way, and to have their turn of audience.[8]

Such a restricted "presence chamber" or mind's eye might seem limiting, but instead, I argue, it has a critical role in keeping within feasible bounds the complexity of the learning tasks that our cognitive system needs to solve.

In a conventional computer we have a very small fraction of the overall hardware investment devoted to the registers, where information is placed that is to be operated on and changed. The remaining much larger fraction of the hardware either stores information or moves it around. In a parallel computer there will be replication of processors and registers, but the fraction of investment in registers, as compared with communication and memory capabilities, is still small.

In biological brains the working memory is probably not as localized physically as are the registers in present-day computers. However, it is believed that working memory at any one time contains much less information than the total contents of the long-term memory. The number of visual concepts that a human may be capable of recognizing has been estimated, by counting the relevant words in a dictionary, to be around 30,000.[9] Total human long-term memory capacity is presumably much larger than this since not all concepts are visual, and since we can recall specific facts and events as well as concepts. Expert knowledge has been estimated to be much larger, perhaps by a factor of ten or more. In contrast, recall that George Miller estimated the maximum number of distinct entities that can be represented in short-term memory as being seven, plus or minus two.

The computational reasons for the amount of information that can be stored in computer registers being small are of two kinds. First, the circuitry needed to perform operations on the registers may be complicated, more complicated than needed for storage and communication. Second, if operations are performed on many registers at the same time, then some systematic way is needed for organizing the cacophony of results that emerge—which is the problem of parallel computation.

Both registers in computers and working memory in brains bring together pieces of information in new combinations, to get results that may never have been computed before. In a computer we may wish to multiply two numbers retrieved from the computer's memory. In the working memory of a brain we may wish to predict the consequences of a novel combination of actions, to see, for example, whether that combination has a promising enough outcome for us to justify doing those actions the next day. In order to predict these consequences a variety of related pieces of knowledge may have to be retrieved from long-term memory.

The computational challenges that arise for registers in computers arise equally in biology.[10] The circuits needed to maintain the working memory may be complicated, as may be also the task of coordinating all that is happening. All this puts a ceiling on how much information the mind's eye can reasonably handle at any time.

I believe that these computational imperatives for having a small working memory, however constraining they may be, are by no means the ultimately limiting ones. Even more severe constraints are imposed by the fact that the brain does not merely have to compute, but also needs to learn. The smallness of the field of view of the mind's eye is essential to make the world learnable. The more information we attend to at a time, the more complex is the task of abstracting regularities from it. Apparently seven (plus or minus two) strikes a useful balance between scope and efficiency. Having our consciousness streamed through such a small aperture serves the function of permitting learning.

Having a small mind's eye forces us to look at the world effectively through no more than a pinhole. Between the world and its enormous complexities and our memories with their highly complex contents is placed this very limited field of attention. As a result, we are forced to manipulate this limited field with some care. We make choices about where to cast our gaze next, what to think of next, and what knowledge from our long-term memories to bring to bear on our thoughts. Making these choices is challenging since, as we have to presume, they will also be based on the restricted amount of information available in our mind's eye.

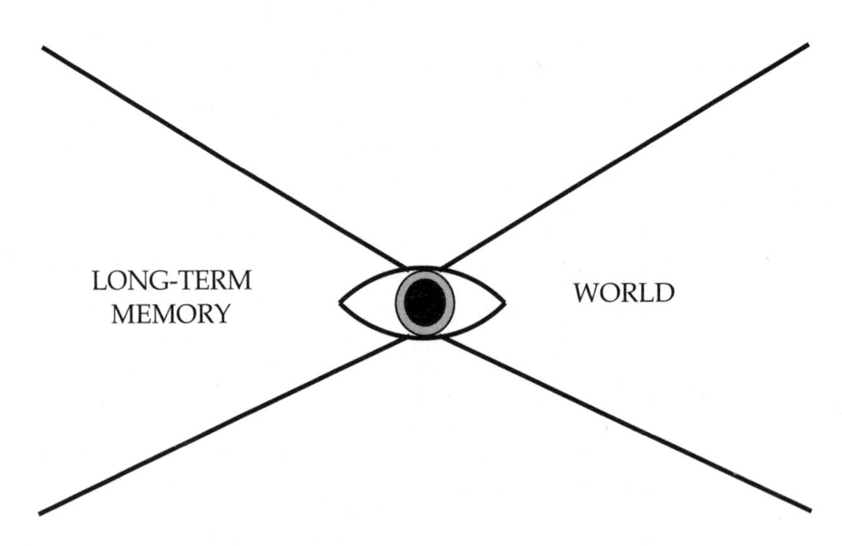

Figure 7.1 The mind's eye is shown occupying a metaphorical pinhole between two funnels, one facing the world, the other the individual's long-term memory. We regard the mind's eye as a computational device that contains the information of which an individual is conscious at any time. The basis of learning is the data that streams through the mind's eye.

The mind's eye therefore can be viewed as the focus of an information funnel between the world and the thinker. It is a two-way funnel that restricts information flow from the outside world as well as from the long-term memory. It corresponds roughly to information we are conscious of, but may include more. It summarizes each experience succinctly enough so as to make both computation on it and learning from it tractable. This succinct summary is informed by both the external input and the internal long-term memory. The succinctness of the description is important for addressing complexity. By permitting learning, it also addresses brittleness.

But most crucially, the mind's eye addresses both semantics and grounding by defining the arena to which all learned knowledge refers: our learned knowledge is derived from real-world experience, but only after filtering, and only to the extent that it is ever represented in the mind's eye. Our learned generalizations have validity (in the PAC sense) for the distribution of the contents of our mind's eye that is generated as we go through our experiences. This then is the semantics and grounding we ascribe to cognition.

7.8 Robust Logic: Reasoning in an Unknowable World

We are approaching the goal of describing the system of robust logic that addresses the four challenges to reasoning. To review, any such system that is to model cognition needs to satisfy two requirements:

(i) All the learning and reasoning processes need to be computationally feasible, in the sense of being polynomial time in the appropriate parameters. The learning process needs to be robust in the PAC sense, as opposed to being brittle, so that any errors in the knowledge can be reduced after sufficient further exposure to the environment to which the knowledge refers. The learned knowledge needs to have clear semantics and grounding.

(ii) Reasoning needs to have a principled basis, in the sense that if two pieces of knowledge each having some PAC accuracy guarantees are applied in succession, then any conclusion so derived should inherit some accuracy guarantees also.

To address these requirements, we have the mind's eye, which we shall now discuss a little more formally than we have so far. Let us call the contents of the mind's eye at an instant a scene. A scene contains a fixed number, say twenty, of undifferentiated tokens, denoted by $t_1, \ldots, t_{20}$, and there is a fixed set of relations that may hold for various subsets of the tokens in a scene.

Each token can be associated temporarily with anything that our mind's eye is then contemplating. The relations come from a fixed set that the system knows about at the time. Suppose that in a particular scene the relation "elephant" is true for t_1, the relation "peanuts" is true for t_2, and the relation "likes" is true for the pair t_1, t_2, in that order, meaning that the elephant represented likes peanuts. These three relations applied to these tokens would represent the contents of the mind's eye at one instant, as illustrated on the left-hand side of Figure 7.2.

Robust logic has mechanisms for learning and reasoning. The novelty is that learning and reasoning will be based on the *same* semantics, in particular PAC semantics. That is the key.

In classical logic a rule would be written as

$$\forall t_1\ \forall t_2\ \text{elephant}(t_1)\ \textbf{and}\ \text{likes}(t_1, t_2) \rightarrow \text{peanuts}(t_2).$$

The ∀ symbol again means "for all." The statement would be interpreted to mean that for any two things t_1 and t_2, if the first is an elephant, and the first likes the second, then it follows that the second is peanuts. In other words, if an elephant likes something then that thing is peanuts.

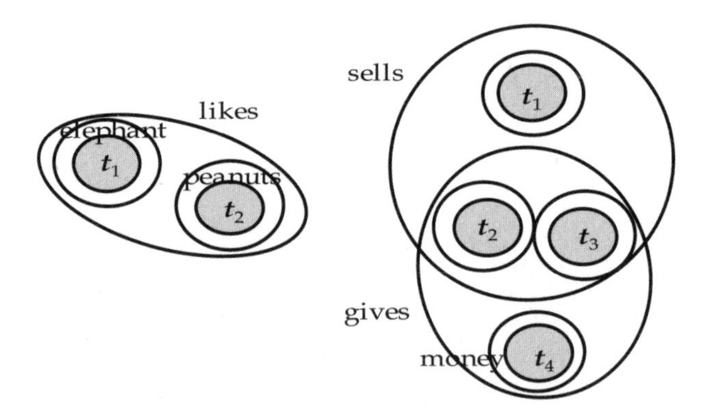

Figure 7.2 Two scenes. The left-hand panel is discussed in the text. The right-hand panel is a little more complicated, and shows a transaction where a person t_2 sells an object t_1 to another person t_3 and is given money t_4 by t_3 in return.

Unfortunately, this kind of semantics is somewhat alien to PAC learning. One reason is that this logical statement is only a one-way implication. It does not intend to imply that being a peanut necessitates that all elephants like it. However, when learning a concept, we want a two-way implication—the learned concept should be a good approximation of the target concepts both when the concept is true and also when it is false. For this reason we shall need to learn rules of the form

$$\text{“complicated condition”} \equiv \text{peanuts}(t_2),$$

where the ≡ symbol denotes such an equivalence, or two-way implication. Note that these equivalences can be used for reasoning in exactly the same way as the logical implications. Whenever the left-hand side is satisfied, we can deduce that the right-hand-side assertion peanuts(t_2) also holds, at least probably.

Now one advantage that machine learning offers is that the "complicated condition" in rules like the one above can be arbitrarily intricate, as long as it can be acquired by learning. In that spirit, we shall allow rules of the general form

$$F([\exists t_1 \text{ elephant}(t_1) \textbf{ and } \text{likes}(t_1, t_2)], \ldots.) \equiv \text{peanuts}(t_2),$$

where the function F is from a class C of functions that is PAC learnable and its arguments are certain restricted expressions to be described later. In this instance the first argument of F asserts that there exists (∃) something (t_1) that is an elephant and likes (t_2). The intention of such a rule is to *predict* for any particular scene whether the set of tokens named on the right-hand side (just t_2 in this example) has certain properties, such as that of being peanuts.

For each word variant in the dictionary, for example peanuts, we can imagine having a rule with that word variant on the right-hand side. The left-hand side will amount to an approximate definition in terms of other words. More concretely, the left-hand side of each rule will express a criterion on scenes for the concept on the right-hand side to hold. The left-hand side may be very complicated and would typically enumerate all common conditions to each of which the answer "peanuts" is a reliable one. In this example one of the many such conditions may be "What to say if asked what elephants like to eat."

In rules of this general form, the left-hand side will be learned from examples in the PAC sense. The learned function F may be complicated—the conditions that provide evidence for something being a peanut may be multitudinous and complex. No human needs to be consciously able to describe it. All that matters is that the function F be from a learnable class C, in which case even large ugly expressions that capture all common forms of evidence of peanuts can be learned, both by computers and brains. It is not implausible that our brains are full of such circuits. After all, it is important for humans to take a position very fast, in a few hundred milliseconds, on whether what we are seeing is a peanut or a tiger.[11]

Robust logic is defined so that rules can be learned and reasoned with in polynomial time. A persuasive candidate for the learnable class C appears to be the following. The left-hand sides are defined to be the class of functions that can be expressed as linear separators, so as to be learnable, but where the variables now are independently quantified expressions (IQEs) such as "$\exists t_1$ elephant(t_1) **and** likes(t_1, t_2)." This last example is an instance of a *schema* that consists of one $\exists$ ("there exists") symbol and is the conjunction of two relations, one with one argument and the other with two, with one token shared between the two and the other quantified. Put a different way, we obtain other members of the same schema from this instance by substituting any other relations for elephant( ) and likes(,), provided they have the right number of arguments.

IQEs have the following two contradictory aspects. On the one hand, they are quite powerful in being able to express complex relationships among objects in a scene. On the other hand, they are simple in that given a specific scene, such an IQE will be either true or false for that scene, and it is easy to determine which one is the case. For that reason we can treat each IQE as a Boolean variable that for any example scene takes value either 0 or 1. In this way we can treat IQEs as features in a standard PAC learning setting and use whatever learning algorithm we like. For example, we can interpret the Boolean values 0 and 1 as numbers and use the perceptron algorithm with these IQEs as features.

We can have IQEs based on more general schemas than the one illustrated, with, say, three or four relations rather than two. However, that could incur much higher computational costs in learning and reasoning. In particular, if we use a schema but have no information about which of the

IQEs defined by it are relevant, then we have to entertain them all during learning. This consideration limits the size of the schemas that are useful in practice.

Where we are heading here is the following. Learning will be done by a conventional learning algorithm, such as the perceptron algorithm, but the variables will now be these IQEs. Reasoning on any one scene will be done by invoking any rules whose left-hand sides are satisfied by the scene, updating the scene with the relations in the right-hand sides of those rules, and repeating this process as appropriate.

The reader might be asking by now: But what does this notation mean? What is the quantification over? Does the ∃ mean "there exists somewhere in the universe?" Does the ∀ mean "for every object on Earth?" No. Either would violate our desire for grounding. The symbol ∃ means simply existence in the one scene in question, and ∀ means universal for every object in that one scene. In other words, to make the IQE "$\exists t_1$ elephant(t_1) **and** likes(t_1, t_2)" true for token t_2 in a particular scene, there must exist some token t_1 in that scene such that the relations elephant(t_1) and likes(t_1, t_2) hold in the scene. Similarly, the ∀ symbol denotes all objects in the scene, not the universe.

In general, several variables can be quantified, some existentially (∃), and some universally (∀). However, the quantifiers in the different IQEs have to be interpreted independently of each other. This means that one cannot assert *directly* that there exists one token that satisfies two IQEs simultaneously. If one wants to assert that, then one has to extend the allowed schema to allow the needed combination of the original two IQEs as a single IQE of possibly double the size, and to accept the greater computational costs that would follow.

These definitions are construed so that, given the left-hand side of a rule and a particular scene, one can evaluate which IQEs are true and which are false for that scene, and hence determine whether the left-hand side holds for that scene. In other words, one can determine whether a rule applies to a scene from information in that scene. No other knowledge is required.

Once we are in this position, we can use any conventional learning algorithm for the chosen learnable class. If it is the class of linear separators, then we may use the perceptron algorithm. Our method of generating all possible IQEs for a fixed schema (i.e., by replacing elephant with any of the other words allowed) creates large but still polynomial numbers of IQEs.

It is important that the number of examples needed be not excessively dependent on the number of such IQEs, most of which can be expected to be irrelevant for any one natural concept. Fortunately, as mentioned in Section 5.11, there exist learning algorithms that are attribute-efficient in the sense that the presence of large numbers of features that are irrelevant does not incur inordinate costs. In particular, the Winnow algorithm can learn disjunctions, using a number of examples that grows only logarithmically, rather than linearly, with the number of irrelevant features.[12]

We must not lose sight of the fact that the purpose of the rules is to make predictions on examples not previously seen. Given a scene that specifies some relations on the tokens, we would like to be able to predict what other relations hold also. In robust logic we shall do this by applying those rules whose left-hand sides hold for that scene. We deduce that the relations on the right-hand side will then also hold for particular tokens in the scene, at least with high probability. We can invoke several rules in sequence in a chain in order to successively fill out the picture of what is implied by the given information. If one is told simply that it is raining outside, one can make several deductions about what else is probably implied using one's database of common sense knowledge. Robust logic is formulated so as to make such predictions have some guarantees on their accuracy.

The technical content of the robust logic system can be summarized as follows. Scenes, which formalize the contents of the mind's eye, occur according to a natural probability distribution that reflects all the complexities of the world, as perceived by the individual to whom that mind's eye belongs. This distribution can be arbitrarily complex, and the individual need not know anything about it. Nevertheless, rules can be learned from examples drawn from that distribution that will be reliable on new examples from the same distribution. (The possibility of this kind of learning is the main content of computational learning theory, which we have described in Chapter 5. Even in a world too complex to fully describe, rules that express learnable regularities can be acquired.) The added content of robust logic over pure learning theory is that it provides a way of reasoning by chaining the learned rules, so that whenever the constituent rules have some guarantee of accuracy, so will also the predictions made. All this is achieved with computationally feasible resources and with controlled error.[13]

As compared to the standard machine learning framework, extra complexity burdens are imposed by the fact that robust logic manipulates multi-object scenes (i.e., the mind's eye contains multiple tokens—for instance, one for an elephant and a second for peanuts). It is this complication that necessitates the use of IQEs for keeping the computations feasible.

7.9 Thinking

Whether we use robust logic as a basis for a computer implementation of a "thinking machine" or as a basis of thinking about human cognition, we need to ask more concretely how the information in the mind's eye—corresponding to Galton's notion of "ideas . . . within my full consciousness," as we quoted him earlier—are to be manipulated.

First, how are the contents of working memory examined and acted on in any such system? In the first instance the answer is simply the following. Given a particular scene, the learned rules in long-term memory are examined, and those whose left-hand sides satisfy the scene and therefore apply to it are selected. The scene is then modified so that the predictions made by the right-hand sides are added to the scene.

Second, how are new contents brought into working memory chosen? For example, when making a purchasing decision, it may be useful to assign a token for "price," if there is not one already, so that in reasoning about whether to make the purchase that notion could enter into the judgment. For this we may extend the syntax of the rules to allow a new kind of operator on the right-hand side for co-opting tokens not currently employed in the scene and imposing the newly desired relations on them.

These are just two components of how such a mind's eye might be managed. There are several other aspects that may also be critical to its effectiveness. For example, we also need mechanisms for deciding which tokens to reassign to new roles as the system moves on with its thoughts. Any system that incorporates robust logic will need some management architecture that encapsulates some such policies.

For learning and evolution I claimed in previous chapters that there were computational models that are robust under variation. I do not have a basis for making this claim for the phenomenon of intelligence as we generally understand it. Rather, I claim that any understanding of intelligence will need to take a principled view of the criteria I enumerated for robust logic, and that robust logic suggests a principled and feasible way of realizing them.[14] While there may be many kinds of intelligence, some minimum ability to reason from learned information, with all the uncertainties that that entails, has to have a role. Furthermore, any reasoning system for the theoryless, including the human system, will suffer from the same frailties as does our robust logic. Chaining together beliefs that we believe to be probably approximately correct can be justified, but the conclusions will also be only probably approximately correct, and the longer the chain of reasoning the larger the errors that we shall have to accept.

This concludes our formulation of the three phenomena of learning, evolution, and reasoning from learned data, in terms of ecorithms. Ecorithms comprise only a subset of the computations that Turing universal machines can execute. But up to Turing's time it was this subset that had domination on Earth.

Chapter Eight

Humans as Ecorithms

No, I'm not interested in developing a powerful brain. All I'm after is just a mediocre brain, something like the President of the American Telephone and Telegraph Company.
ATTRIBUTED TO ALAN TURING

8.1 Introduction

Science, whether of nuclear reactions or the impact on health of smoking, does not dictate how it should be applied. Its relevance needs a separate, if theoryless, discussion. Scientists are justified and perhaps obligated to speculate on the broader relevance of their work. This is the excuse that justifies what follows here.

The previous chapters have expounded the thesis that a decisive determinant of life is the ecorithmic relationship between living organisms and their environment—life coping with its environment by means of learning mechanisms. In the remaining chapters I shall try to provide a personal, and alas theoryless, answer to two questions: Can all the complexities of life, intelligence, and culture that we witness on Earth be explained by this hypothesis, and do any consequences of general interest follow from it? Necessarily the discussion will be much more speculative than before.

One difficulty with making this discussion more theoryful is that the fundamental ecorithms used in biology have not yet been identified. A major motivation of our study, of course, is exactly to encourage further work toward filling those gaps. Once these ecorithms are better understood, the topics we are about to discuss will become more amenable to scientific analysis.

In the meantime there is a predicament. Until now I have been discussing theories of computation and learning, which I regard as scientific. As I move on I get into more dangerous territory because I will also be making statements about what is currently theoryless. One of my strongest conclusions will be that one should be extremely skeptical of discussions that are entirely about the theoryless, because little justified meaning can be ascribed to them. I need to be careful not to fall into that intellectual tar pit myself. If I manage to perform this feat, it will be by keeping some science always in sight, in the sense that I try to relate important but theoryless questions to the theory of learning.

8.2 Nature Versus Nurture

A widely debated but still unresolved question is how much of human behavior is determined at conception or birth, and how much by life experience; whether it is that genetic differences among individuals or groups are decisive, or that humans are so adaptable that life experience after birth is all determining.

Definitive answers have proved elusive. Humans are remarkably trainable and educable after birth. Stark differences in performance among individuals and groups on measurable tasks such as math tests are easy enough to find. Experiments or surveys that detect such differences are sometimes highly reproducible. However, it is usually open to interpretation whether the differences found are relevant to this debate.

The nature versus nurture issue would seem an unlikely difficulty for a computational theory. Surely, in an electronic computer there is no such problem. It is clear what your new computer's capabilities are at the moment it is delivered to you. These capabilities ought to be easy to separate from subsequent enhancements, such as added software or hardware, or from running a learning algorithm that learns to recognize your voice.

An ecorithmic viewpoint is different, however, from the caricatured computer just described. Instead, it corresponds to the case in which your computer was itself created with the aid of a learning process before you purchased it, and it learned more and more knowledge during your subsequent ownership. If the learning processes used before and after your purchase were essentially of the same nature, then it may be truly problematic, after the fact, to reconstruct what the machine's capabilities had been at the moment of purchase. What if you had bought a used computer that had been trained by its previous owner in a similar way as you train yours? What if you buy a dog and try to differentiate between the influence of the previous owner and you?

I suggest that the source of difficulty of the nature-versus-nurture problem for humans is exactly this. The evolution process before conception and the learning process afterward are just too similar to necessitate or impose a simple interface at which they can be differentiated. Consider the absurdity of trying to distinguish which parts of a person's character can be attributed to experiences before their 5½th birthday, and which afterward. Similarly, the moments of conception or birth are not overarching watersheds. When we are conceived or are born, we already have many important things half-learned. The question of nature versus nurture fails because it imposes an almost arbitrary instant into a continuum of change.

8.3 Naïveté

Earlier I described the notion of learnable target pursuit, an aggressive strategy for pursuing every possible opportunity for learning that the environment provides. Whichever concepts are computationally learnable with respect to the concepts that are already recognized as features will be learned if the environment provides appropriate examples. I suggested that our cognitive system is wired for doing this.

A system so wired is not without its shortcomings. A creature constantly on the lookout to assimilate any and all learnable regularities is highly fallible. The pitfalls are the same as when interpreting statistical information generally—the data may be unrepresentative or insufficient, or statistical correlation may encourage us to assume unwarranted causation.[1] The adage that "there are three kinds of lies: lies, damned lies, and statistics" expresses a fundamental trap into which such a cognitive system may easily fall. In particular, our aggressive algorithms for learnable target pursuit will be a little naïve when dealing with unnatural or adversarial sources of data.

To put this in more technical terms, in PAC learning the environment is regarded as neutral. It is neither trying to help us learn faster, as a benevolent teacher would, nor is it adversarial in seeking to confuse or delay our learning. The physical world seems to be essentially neutral in this sense. It is therefore plausible that such a neutral setting for learning is the appropriate one in which to consider how evolution has solved basic physical problems, such as locomotion and vision. No benevolent source was apparently available to speed up learning, and no adversarial one to impede it.

Our primary learning instincts will then have evolved to respond to the world as if it were neutral, and to take at face value and as typical all the information found therein. As we have seen, a benevolent person, such as a parent or teacher, can take advantage of this situation by presenting us with information that drives our hard-wired learning algorithms to the desired generalization especially quickly. On the other hand, an adversarial entity can mislead us, by presenting information that will quickly lead our algorithm to a false generalization.

When we meet a new person, we are inclined to believe that the behavior we see is typical of their behavior and not a deceitful act. When we go to a restaurant, we are inclined to believe that we are getting a meal typical of that restaurant. The unrelenting pursuit of learnable targets and the predisposition to view the masses of information we process as neutral make us easy victims of misleading coincidences or deceit.

This phenomenon is pushed to the extreme when we learn from a single example. It is reasonable for a child to be trusting when learning concepts from books and not to be constantly on guard that the examples shown, of elephants or whatever, are unrepresentative. Yet with this over-eagerness to take data as representative, we may easily be driven to wrong conclusions with the smallest of manipulations. One negative word dropped about a person we little know may influence our view of that individual permanently.

8.4 Prejudice and Rush to Judgment

A related pitfall arises from the fact that we sometimes have to make decisions fast. Having a circuit that in any situation tells us instantly whether or not to run away is vital. Although the output of such a circuit could provide information for further extended deliberation at some moment of ease, its primary function will be a yes/no recommendation as to whether to run in some stressful situation. Unfortunately, the need to quickly decide whether or not to act may give our circuits a tendency to rush to judgment on scant information. By scant information I mean that it would be insufficient to provide a confident prediction of the future even in a neutral world. This is not to say that scant information never warrants action, however; it certainly does if the alternative would have grave consequences. When walking in grizzly country, one should take note of growling in the undergrowth.

There may be additional reasons why learning algorithms may take a position long before there is overwhelming evidence. The perceptron algorithm we described earlier was first proposed because its step-by-step processes appear well suited to neural systems. It has some additional properties that are relevant here. In current terminology we would call it an online algorithm in the sense that after any number of examples, as few as one or even none, it takes a position on the next unlabeled example, classifying it as positive or negative. If our brains have such online algorithms wired in, which I believe they have, then at any instant they will have a tendency to take positions on every question. In other words, they are wired to rush to judgment.

Rushing to judgment may be to our overall benefit in a neutral world. We no doubt make strong judgments of individuals after one meeting, and of restaurants after one visit. This eagerness to prejudge based on scant evidence is, I believe, a fundamental aspect of our nature, a consequence of having a decision-making system capable of responding to even the least information. We are better at making decisions than in evaluating the probability of their correctness or in justifying our actions in terms of the evidence. This would reflect the relative ease of PAC learning yes/no questions, as opposed to these more complex tasks.

We may, in fact, carry this trust even further. Besides making judgments based on our own experience, we may also be unreasonably receptive to adopting the beliefs and prejudices of others, perhaps assuming that they, or perhaps a third party, had acquired the beliefs in question from data just as we would have. This behavior may be an effective strategy in a world that allows facts and statistics to be taken at face value, but it is a severe weakness in the presence of adversaries and entities with agendas.

Recently I got into a taxi in Philadelphia and within a minute the driver asked: "Are you a mathematician?" I said: "Why?" He explained: "Mathematicians either have no hair or their hair sticks up." Apparently he had studied some mathematics at university. At least his prejudice on this matter did have some basis in personal experience.

8.5 Personalized Truth

I have emphasized that PAC learning offers a theory of how people all over the world can arrive at similar notions from different personal experiences. The explanation is that as long as humans have a shared learning algorithm, and as long as the examples they see are of the same concept, then each person can learn from examples labeled by others who had learned that concept previously. It is not necessary that the actual examples seen by individuals around the world be the same. However, PAC learning has some characteristics that impose some evident limitations also. In different parts of the world the same word may have different meanings, or the distribution of examples may be different. In these cases the Invariance Assumption would be violated, and shared meaning would not be achieved. Misunderstandings would result.

There are pernicious obstacles to shared meaning even beyond those inherent in differences in meaning and distributions. These further impediments are imposed by the constraint of a limited mind's eye interacting with an internal memory full of beliefs. We may all be looking at the same world through our mind's eyes, but since we have much control of what information to allow in, dependent on our beliefs, we may not all *see* the same world.

In the mind's eye we process not only the information coming from the outside, but also information internally retrieved from our long-term memory. When we listen to a political debate, we can each bring different internal information to bear to our mind's eye that will color the way we interpret the external input, and even choose different sentences from the input to acknowledge or ignore. Differences in our past experiences are not necessarily ironed out by a common present. The choice of which experiences to process, and how, are personalized according to the contents of long-term memory.

The observation that different individuals see the world differently is no more than a truism. What is offered here is a view of why it has to be this way: We must view the world through a limited mind's eye—no more than a pinhole between the large world of our long-term memory, and the even vaster world outside—if the world is to be computationally learnable. That requirement has resulted in our having control of what information from the outside world we allow to impinge on our mind's eye—through our decisions of where to go, whom to believe, and what to look at—as well as what constructions and beliefs from our internal experience we project onto it.

Our long-term memory is filled with material filtered through our mind's eye. The generalizations we have learned work well in making useful predictions only for scenes typically encountered by our own mind's eye. A cognitive system that uses PAC semantics does need to achieve approximate internal consistency, and no doubt does so. Hence consistency of concepts within one person is to be expected as a norm, and this may be viewed as personalized truth for that person. But this does not imply that there will be consistency of judgment among different individuals.

8.6 Personal Feelings

Humans use words such as love, hate, pride, and guilt to describe feelings. These words evoke understanding and responses from others. Words written by one generation speak to another. There is much shared meaning.

Personal feelings and opinions have reality, both in the brain circuits and in the actions of individuals. Talking about our opinions and feelings might be pleasurable experiences and may be so justified. But what other status can we ascribe to them besides the fact that they are real? The ecorithmic position on this is simply that, however real or strongly felt personal feelings and opinions may be, they have no status beyond merely being the outputs of circuits that compute the theoryless.

Given that humans are biologically almost identical, it is only to be expected that we have similar experiences of pleasure and pain. If the circuits in our brains are similar, then it is not surprising that if we report pleasure or pain given an experience, then other individuals will understand us simply because their circuits are similar. It is not surprising then that there would be widely shared notions regarding basic ethics. Of course we are not exactly identical, and hence these notions may not be quite the same for different individuals. But the idea that we have some shared notions of pleasure, pain, and ethics is only to be expected. Both the reality of personal feelings as well as the possibility of communicating about them with others are entirely consistent with the ecorithmic viewpoint.

Whether we can reason about our opinions and feelings, shared or otherwise, is another matter entirely. I know of no basis on which we might justify applying any such process and reaching certainty in any such theoryless area. Individually, we may be as justified in fighting for our opinions and feelings, and acting on them, as anyone else. But we should do both with caution and humility, since there seems no way of justifying the superiority of our feelings and opinions over those of others.

8.7 Delusions of Reason

A further consequence of one of the arguments in this book—namely, that when reasoning about theoryless concepts, the semantics of PAC learning is the one that is appropriate—is that reason has its own limitations that are ignored at one's peril.

Words in natural languages are applied in practice to naturally arising situations. They owe their usefulness to their being understood consistently enough when used in their conventional senses. For example, words such as "conscious" refer, in the first instance, to personal experience and feelings. One has to be struck that the level of consensus in how such words are understood is as high as it is.

If we try to apply these words to thought experiments or artificial situations divorced from the natural contexts in which they had been learned, we enter the realm of the meaningless. Is a table glued to two chairs a table or a chair? Is a termite conscious? Can a computer have free will? Such questions are meaningless because the concepts the words describe have meanings only for the distribution of examples from which they have been PAC learned. No standing can be ascribed to them beyond that. Thus, the widely discussed question of whether a computer has consciousness if it faithfully simulates the changes in the brain of a human is of the same status as the question of whether an animal with two wings and a tusk is an elephant or a bird. There is no reason why we should pay any attention to it. It is simply fallacious to apply PAC learned concepts, such as consciousness, to artificial situations that do not occur in the domain from which the concepts were learned.

Unfortunately, standard logic, when applied to propositions expressed in English or any other natural language, offers no end of opportunities for committing such a fallacy. This is particularly the case in philosophical discourse in which one attempts to gain insight by discussing thought experiments that invoke unusual or marginal situations. For example, some counterarguments to the possibility of artificial intelligence have the following form. A computer program with one instruction cannot possibly be considered conscious. But if we suppose that humans are each equivalent to a million-line program, say, and are conscious, then there must exist a minimum number of lines that qualifies for consciousness. Any specific such number is then argued to imply an absurdity.

As long as the notion of consciousness is used in the conventional informal sense, this argument is again meaningless. It may be true that, with respect to the experiences that a human associates with consciousness, no one-line program is conscious. It may also be true that human brains can be faithfully simulated by a million-line program. However, it does not follow that the human experience of consciousness can be used to classify any program of intermediate length as being one or the other. For that reason we would consider this whole line of argument to be fallacious—it wrongly assumes that the notion of consciousness is theoryful in a sense in which it is not. (Of course, this is not the only way in which one might find fault with the argument.)

An analogy in the realm of character recognition would be the problem of distinguishing between handwritten examples of the numerals 2 and 3. We may be highly confident in natural situations when classifying whether someone has written a numeral 2 versus a numeral 3. But it is easy enough to write a deliberately ambiguous figure intermediate between those numerals, designed to split the opinion of reasonable people equally. For such figures it does not make sense even to ask the question.

Figure 8.1 Are these 2s or 3s?

The misplaced illusion of having reason on one's side can arise not only in philosophical discussions of hypothetical situations. In all areas that are theoryless, including many of substantial everyday human concern, such as politics and religion, we are faced with the same dangers. We may be on safe ground when reporting on observable facts, and when interpreting statistical evidence. We are also individually competent in handling a wide range of learned theoryless generalizations in an internally consistent way. The problem arises when we go beyond this. If we attempt to apply reason by putting together two theoryless ideas, the robust logic described in the previous chapter offers only a probably approximate justification. I do not know of any more certain source of justification. Treating as certain any predictions about the theoryless, whether made via robust logic or any other known system, would be fallacious.

There are dangers in believing that something is theoryful when it is no more than an expression of personal feelings. Political belief systems have transformed nations, but when their time has passed it is difficult to reconstruct how they could have gripped the imaginations of so many. Similar dangers arise in areas that look more quantitative and technical at first sight. With hindsight, the various models of risk that dominated in the financial industry prior to 2008 were not theoryful, but similarly only expressions of personal feelings.

A necessary, but not sufficient, criterion for a field being theoryful, and appropriate for logical reasoning, is that substantial consensus among informed humans can be reached regarding it. Currently in many fields of much human significance no such consensus is in sight. Discourse in these areas may have value as communication of facts, as exchanges of personal views, or even as entertainment. The question arises, however, whether attempts at reasoning or other discourse in these fields have any further value.

Perhaps one of the most extreme expressions of confidence in reasoning is the description of intellectual developments in Europe in the seventeenth and eighteenth centuries as the "Age of Reason." This seems curious nomenclature, given that in that era, humanity did not appear to increase its ability to reason. The main development was that some areas, such as mechanics, were brought within the scope of science and the theoryful. Within those areas the application of reason proved powerful and effective, indeed, and those successes in science did inspire philosophers and others to attempt to apply reason to the areas that had remained theoryless. But whether these latter developments can be treated also as triumphs of reason seems open to debate.

8.8 Machine-Aided Humans

The circumstances that led to the financial disasters of 2008 point us toward a second consideration—namely, how to approach the combined activity of human and computer. Financial and investment analysis is one area in which statistical tools are used extensively to analyze data, and this is also true in many areas of the social sciences. These statistical tools, often in the form of packages of computer programs, act as artificial enhancements to the human ability to spot patterns in data. While the statistics of the data that the machine computes may be influencing the conclusions drawn, it is still the human who draws the conclusions. Hence, in such cases the combined activity of the human and computer may be viewed still as an ecorithm.

With this view, we can regard the use of statistical algorithms as very natural extensions of what our biological circuits do in learning. Such activities are laudable enough, enhancing as they do the powers of our natural ecorithms for coping with the theoryless. However, the caution that is appropriate for any theoryless decision making remains even if it is machine enhanced. The resulting decisions are still likely to be theoryless. Enhanced statistical tools do not make a theoryless domain theoryful, as both economists and financiers have cause to appreciate. Any decisions that come out of this process will need to be treated with the same caution. Reasoning about any findings that are made will have the same limitations that we have discussed for any theoryless decisions. Unless there is evidence to the contrary, the conclusions drawn from statistical analyses should be viewed simply as the opinions of the analyst, augmented as the analyst's circuits may be by artificial means.

I am not arguing that the most advanced theoretical methodologies known should not be employed, but only that their limits be fully acknowledged. In areas of the theoryless, including the social sciences, human decision making, even when augmented by the most elaborate intellectual aids, is still subject to the uncertainties that are inherent in PAC learning.

8.9 Is There Something More?

Earlier chapters had presented ecorithms as a mathematical field of enquiry, and suggested that it offered an unevadable key to understanding the phenomena of learning, evolution, and intelligence. In this chapter I have reviewed some consequences of this suggestion. The final question that remains is whether this approach is missing something fundamental with respect to the human mind.

The view of cognition given here may be summarized as follows. Cognitive concepts are computational in that they have to be acquired by some kind of algorithmic learning process, before or after birth. Cognitive concepts are equally statistical in that the learning process draws its basic validity from statistical evidence—the more evidence we see for something, the more confident we will be in it. There is an interface—our mind's eye—between the world and our long-term memories. We, as represented by our circuits, have some control of the information flow through this mind's eye, which therefore cannot be viewed as objective or neutral. The circuits in our nervous systems that comprise our database of knowledge are the results of many successive acts of evolution and learning. In other words, our knowledge is accumulated, to echo Darwin, in a series of successive, slight modifications, each one having the semantics of learning. An instance of reasoning consists of applying these circuits, within the mind's eye, to whatever situation is at hand. Our neural systems evolved to cope with the theoryless, but we use the same circuits when we make theoryful decisions.

Can the human mind be based on something quite as simple as this picture suggests? I argue that the answer is yes, that there is little evidence that in order to explain human behavior mechanisms of a radically different nature are needed. Those who insist otherwise may have an overly optimistic view of the capabilities humans actually possess. When new problems arise, we are not that impressive at solving them. Any solutions found can usually be traced to solutions to similar problems encountered earlier. Were the responses to the financial crisis of 2008 impressive as leaps of the human intellect? I, and many others, would argue not. The responses implemented were variants of responses used during previous similar episodes. They may have been ultimately effective, but they do not offer evidence of an out-of-this-world quality of mind. Rather, they confirm that in order to tackle complex theoryless problems learning from the past is as good as we know.

The reader may choose to look to other human activities to find evidence of cognitive mechanisms of an entirely different quality. I am skeptical that such evidence can be found. This book celebrates science. The history of scientific discovery reveals many brilliant moments. Can we find evidence there of the insufficiency of ecorithms for explaining our intellects? I suspect not. As I have pointed out, science has stunning unities within it. Physical scientists have for centuries been exploring mathematical formulae that express predictive physical laws. There have been plenty of such formulae to find, and they have shared many similarities. Here I have been suggesting that the somewhat different direction taken by Turing will offer the path for understanding a whole host of other currently ill-understood phenomena.

Indeed, as impressive as it may initially seem, the history of science gives us only limited cause for self-congratulation. Newton, Darwin, Einstein, and Turing posed new questions and pursued them relentlessly. While one can only marvel at the insightfulness of their choice of problems and the intellect and persistence they applied, their success would not have been possible without the yet-to-be-explained unities that pervade their science. Even as the universe seems a neutral teacher, we benefit from its unities, which make the task of scientific discovery more tractable than we in our vanity might like to think.

Chapter Nine

Machines as Ecorithms

Why is artificial intelligence difficult to achieve?

> *It is customary, in a talk or article on this subject, to offer a grain of comfort, in the form of a statement that some particularly human characteristic could never be imitated by a machine. It might for instance be said that no machine could write good English, or that it could not be influenced by sex-appeal or smoke a pipe. I cannot offer any such comfort, for I believe that no such bounds can be set. But I certainly hope and believe that no great efforts will be put into making machines with the most distinctively human, but non-intellectual characteristics such as the shape of the human body; it appears to me to be quite futile to make such attempts and their results would have something like the unpleasant quality of artificial flowers. Attempts to produce a thinking machine seem to me to be in a different category. The whole thinking process is still rather mysterious to us, but I believe that the attempt to make a thinking machine will help us greatly in finding out how we think ourselves.*
>
> Alan Turing

9.1 Introduction

I have always had some disquiet about the term "artificial intelligence" and only rarely identified myself as working primarily in that area. However, I remember the first time I met Edsger Dijkstra. He was noted not only for his pioneering contributions to computer science, but also for having strong opinions and a stinging wit. He asked me what I was working on. Perhaps just to provoke a memorable exchange I said, "AI." To that he immediately responded, "Why don't you work on I?"

He was right, of course, that if "I" is more general than "AI," one should work on the more general problem, especially if it is the one that is the natural phenomenon, which in this case it is. In fact, that is just what I believed I had been doing all along. In this book essentially everything said up to this point that applies to artificial intelligence also applies to intelligence. But we all know that intelligence in its natural and artificial forms are not necessarily the same. This chapter discusses the difference.

Previous chapters have sought a computational view of learning and intelligence that is explicit enough that there would be no inherent problem in simulating either by computer. So, you probably feel justified asking, what has stopped us?

One problem is that, although the practice of machine learning lends credence to the notion that learning is central to intelligence, the learning algorithms that are hard-wired in the human brain are yet to be identified. Nevertheless, machine learning algorithms originally inspired by their natural counterparts are already in widespread use, and useful regularities are being found by existing machine learning algorithms in all kinds of data, whether generated artificially or by nature. More broadly, although the ecorithmic approach sheds light on the problem of emulating human intelligence by machine—it suggests that a learning-centered approach needs to be used in the effort—it also points to some serious obstacles. One of the chief goals of this chapter is to explain how the ecorithmic view gives a persuasive explanation of why artificial intelligence is proving so much more difficult to achieve than expected. This may sound like an excuse, but no one can deny that a good excuse is badly needed.

9.2 Machine Learning

Faced with any task, a computer can either be programmed to handle it, or it can be made to learn how to handle it, or perhaps a combination. As we have seen, learning has a statistical aspect and for that reason cannot be made error-free. Compared with the possibilities of creating a faultless program for a task, the learning solution will always have this essential weakness. In applications where we know how to specify the exact outcome we want, it may be best to simply program it, assuming that we are able to do so.

Of course, we may not be able to program it for a variety of reasons. Learning becomes more powerful, and indeed indispensable, (i) whenever we cannot specify explicitly the outcome we want, (ii) whenever we cannot specify exactly what the system that is to execute it knows already, or (iii) whenever we cannot get direct programming access to the system. When the system in question is a human, all three of these conditions hold, and we have no alternative to learning. When the device is a computer, conditions (i) and (ii) sometimes hold. Learning solutions then become desirable for computers also.

Machine learning is now an important and pervasive technology. Important applications to date have been mostly of the first category, where we do not know how to specify explicitly the desired outcome. One can also foresee applications of the second category, in which, as in robust logic, learned knowledge is added in successive layers, so that the possibility of specifying exactly what the system knows is quickly lost, even if it was there at the beginning.

Important application areas of machine learning are presented by the World Wide Web, where vast amounts of new information are created every day, and there is no hope that any single human, or even a group, can keep pace in understanding it. Detecting spam in email is an example. New sources of spam can appear every day. The labor of following every new kind, and manually incorporating ways of detecting it into email systems, would be prohibitive. Much more effective is the use of machine learning algorithms that learn some regularities that distinguish between email that users label as spam from email which they do not. Other applications enable search engines to rank Web pages, to determine which ones to show at the top of the page in response to a user's search terms. The evidence of which pages people click on, when offered alternatives, is invaluable information. Similarly, with online advertising, learning from the history of click-through rates on different websites provides valuable information about where it is most profitable to place advertisements.

Other important applications include natural language processing and computer vision, in both of which one is attempting to simulate human capabilities. In natural language processing a basic problem is spelling correction. Within a text one may have the words *desert* or *dessert*. Humans can usually determine which one is intended from nearby words, *camel* being suggestive of the former and *meal* the latter. One can attempt to automate this task of spelling correction either by programming it, or by learning. Simple machine learning approaches can predict accurately the correct choice when trained on correct sentences that contain the target words, and this latter approach has been winning in recent years, producing more accurate systems. In the area of computer vision one critical task is to recognize automatically what categories of objects are present in a picture. Here again the current methods of choice involve machine learning algorithms.

The pervasive success of machine learning rests in significant part on the effectiveness of a few recently discovered learning algorithms. One remarkable innovation has been boosting, a generic technique for improving the performance of almost any basic learning algorithm. Its discovery arose from investigations into the robustness of the PAC learning model. PAC learning is defined to require the hypothesis to predict with arbitrarily small error. In the same spirit, one can also define the notion of weak learning, which instead has the weaker requirement that the hypothesis predict just detectably better than random guessing.[1] This would be a useful form of learning for gamblers for whom it is sufficient to predict a little above the odds. One can define what it means for a class of concepts to be weakly learnable in this sense, and again one insists that learning succeed for every distribution since, as previously noted, learning algorithms that are specialized to a narrow range of distributions are not too useful.

Somewhat astonishingly, it can be shown that any such class that can be weakly learned for any distribution can also be learned for every distribution in the strong standard sense! In fact, one can automatically translate a weak learning algorithm into a strong learning algorithm. The idea is to use the weak learning method several times to get a succession of hypotheses, each one refocused on the examples that the previous ones found difficult and misclassified. The fact that the weak learning algorithm works for any distribution permits one to modify the distribution at each stage so as to achieve this repeated refocusing. Note, however, that it is not obvious at all how this can be done, since when we are presented with a new example to classify, we have no immediate way of knowing whether the previous hypotheses would have got it right or wrong, let alone whether they would have found it difficult. Nevertheless, as Robert Schapire showed in his PhD thesis in 1990, this can be done.[2] Further, in subsequent collaboration, he and Yoav Freund found a very simple and efficient way of achieving this, called Adaboost.[3] Their method is widely used with a variety of standard learning methods playing the part of the weak learner. Boosting has been found to be a practical and generic way of improving the predictive accuracy of a wide variety of basic learning methods, often even when there is no proof that the methods used are true weak learners.

Besides generic methods such as boosting, a variety of more particular machine learning algorithms are also in widespread use. As we have mentioned before, PAC learning is a specification of what needs to be achieved when learning, and is not ideological about how best to achieve it. An important empirical finding has been that simple algorithms with analyzable behavior often perform remarkably well.

The perceptron algorithm we have described is already a remarkably effective algorithm. As we observed, the restriction to linear functions can be removed by using nonlinear terms as features, but only by incurring correspondingly higher costs in both computation and the number of examples needed. One effective way to have the power of a larger feature set without all the cost, for perceptrons as well as for some related methods, is called the method of kernels.[4] Relative to the straightforward method, kernels can save substantially on computational costs, but not on the number of examples needed. The widely used support vector machine method follows this idea, but it chooses a separator that maximizes the margin (which we discussed in Section 3.7) rather than one chosen by the perceptron algorithm.[5]

For certain problems even simpler methods can be highly effective. An example is the nearest-neighbor method, in which the labeled data items are simply stored and no hypotheses are ever generated. When a new item to be classified arrives, it is compared with the stored items; the one that is nearest according to some criterion is identified, and its label is predicted as the label of the new item. The recent successes of language-translation software can be attributed to the power of this technique. By examining enormous datasets of pairs of corresponding sentences in two different languages, valuable information can be extracted about how a particular new sentence in one language should be translated in the other.

Orthogonal to the question of the choice of actual learning algorithm is the choice of basic features or variables to use. Good choices can yield more accurate predictions. For example, for a computer vision application, starting with the brightness or color of individual pixels as features is always possible but often does not give the best results. Starting instead with some programmed features that depend on a set of neighboring pixels, rather than just one, usually gives better results. The average brightness of a region, for example, may be more informative than the brightness of individual pixels. Biological systems are also known to use such higher level features, which presumably were acquired through evolution, being too difficult to learn from scratch by an individual. In computers the effectiveness of alternative sets of programmed features can be compared experimentally by trying them out. As far as the learning algorithms themselves, there is a small set of methods that are generally effective for the many kinds of data that are encountered. The performance obtained from many of them can be often further enhanced by boosting.

Machine learning has had striking successes for data from almost all sources. One illustrative example is data from the brain. Even though very little is understood about how information is represented or processed in the brain, machine learning algorithms can be trained to make predictions about what a human is thinking on the basis of data recorded from the brain. When a person is viewing a word of text, information about whether the word refers to tools, animals, or buildings, for example, can be recovered from functional MRI images of the person's blood flow in the brain.[6] This recovery is achieved by standard learning algorithms applied to the images as examples and the categories of words as the labels. These images constitute a theoryless arena since we understand so little about how knowledge is represented in the brain. Yet these images apparently abound in regularities that can be learned. This is an excellent illustration of the fact that learnable regularities may be found even in the most complex and theoryless data.

In recent decades machine learning has benefited enormously not only from the discovery of better algorithms, but also from much faster computers and more plentiful data. In fact, the availability of data on scales not before seen is a major new development in our civilization. The potential rewards of data mining on the scales at which it is now possible are immense, and currently largely untapped. The success of machine learning technology on such a broad variety of tasks is strong evidence of the effectiveness of learning algorithms in areas relevant to human information processing. It provides indirect support for the centrality of ecorithms for the even broader range of phenomena that we are claiming here, such as for common sense reasoning.

Can one easily tell in which applications one expects machine learning to succeed, and in which not? A minimum requirement, I would say, is that the distribution on which the system will need to perform well (the *D* in the definition of PAC learning) should be identifiable. This does not mean that one needs to be able to describe it explicitly. Quite to the contrary, it means merely that one can identify it unambiguously. One should, for example, be able to construct a database of typical examples such that if a system works well on that set, then it should work well also "in the field" where success will be determined. If you want a system to play the game of *Jeopardy!*, for example, defining the general knowledge task in vague philosophical terms is not enough. Instead, you will need to identify a distribution of typical questions, perhaps from previous actual games. If you want to build a computer system that can grade student essays well, you need to specify a distribution, say by the language, the topic, and the era in which the essay was written. You have to make some choices. If a dataset of typical instances cannot be created, the prospects of training a computer to do well on the task are slim. If such a database of typical instances can be produced, then the prospects are much better, but not guaranteed. Failure may occur because the needed regularities are inherently hard to learn or beyond current capabilities. Failure may equally occur because the information in the dataset is not sufficient for the task at hand. In the essay-grading case, for example, we may be stymied by the difficulty of acquiring all the common sense knowledge that students know but that is nowhere written down.

9.3 Artificial Intelligence—Where Is the Difficulty?

The quest for machines that approach human competence in handling common sense knowledge has met with many disappointments. If humans manage to achieve this competence, surely it cannot be impossible to replicate. The difficulties, however, appear to be at least substantial.

The source of the difficulty, I believe, is implicit in something we have discussed here already in the context of the nature/nurture debate: the similarities between the two processes of evolution and learning. An individual's cognitive system is the outcome of hundreds of millions of years of evolvable target pursuit, followed by several years of learnable target pursuit after birth. Turing talked about educating a computer as one would educate a baby. This sounds plausible as long as we can manufacture a computer with similar capabilities to a baby. Unfortunately, the state of a baby is the outcome of evolutionary ecorithms and, with respect to our current understanding, is theoryless. There is no reason to believe that the states of babies are easy to describe. The algorithms of evolution may be theoryful, but their results remain, from our perspective, theoryless.

If we had a good theory of which primitive features are computed by the human nervous system at birth, then we could program a computer to compute those primitives, and then perhaps educate the computer just like humans educate babies. If, however, the moment of birth is not a viable starting point for any such course of education, it is not clear whether there is any good place to start. In the worst case there may be no alternative but to start from the beginning of life, billions of years ago, and simulate all of evolution. This would be unfortunate, since the conditions and particular inputs that accompanied evolution over all those billions of years may be just too difficult to determine for us ever to succeed in this.

One can hope for an intermediate solution, where a partial understanding of human functionality at birth, or at some earlier stage of development, is sufficient. This may be our best hope. We would need at least a basic understanding of the functioning of the human vision system, for example, at that stage. There is a lot of evidence that even this system is very subtle and leaves much learning to be done after birth. But even if the visual system is amenable to such intermediate treatment, other domains may prove more difficult. At the cognitive level, babies are more thoroughly prepared for life than was once believed. As indirect evidence, take the difficulty AI researchers have had in formalizing all the common sense knowledge we need to understand everyday life. In order to understand a novel many facts need to be known that are not stated in the novel; many are so obvious that they are nowhere stated in print. This is not merely a feature of complicated adult novels; it has been remarked that children's stories require almost as much common sense knowledge as do novels written for grownups. Unfortunately for Turing's dream, babies arrive miraculously well informed and well prepared to be informed even better.

The question arises as to how humans acquire whatever they learn after birth that is nowhere stated explicitly. Of course, there is a range of possible modes, including vision, smell, taste, and touch, that bypass language. We have previously addressed the issue that such knowledge needs to be treated as learned knowledge so that all its uncertainties can be adequately addressed. Specifying this knowledge in any form, whether in its original mode, or as described in words, is difficult and beyond our current capabilities. The more obvious the knowledge, the more difficult it seems to identify how a human acquires it.

Turing addressed this question in a discussion of the capabilities of a computer that is "a 'brain' which is more or less without a body."[7] He observed that the areas of game playing (such as chess), translation of languages, cryptography, and mathematics were well suited to such a brain because these tasks require "little contact with the outside world"; it was cryptography, he thought, that might prove "most rewarding." This was highly prescient. Our computers encrypt automatically for all of us when communicating, and I at least have taken advantage of computers for the other three activities also. These four areas Turing contrasted with a fifth, language learning: "Of the above fields the learning of languages would be the most impressive, since it is the most human of these activities. This field seems however to depend too much on sense organs and locomotion to be feasible." In these few words he had summarized an essential feature of the problem of acquiring common sense knowledge.

Leaving aside learning processes after birth, what can we say about the problem of characterizing babies at birth? In order to understand the circuits of a baby, one could proceed by behavioral experimentation, observing babies under many different external stimuli to learn what their response is. We would then be hoping to induce what these complex computational systems do from their behavior on various inputs. However, the negative results in learning theory reviewed earlier suggest that this avenue has its own difficulties. Unless the circuits are from a simple enough class, it may be inherently infeasible to perform this induction.

A final, most extreme approach would be to use some yet unknown technology to copy the circuits of a baby cell by cell, or molecule by molecule, faithfully replicating all their functionality. As we saw in Section 3.6, it may be difficult to deduce what a circuit does from its description. So even if this approach becomes feasible one day, it would not necessarily enhance our understanding.

The artificial intelligence challenge has always looked enticing as a scientific problem because of the argument that it attempts only to emulate systems that already exist in nature. It is made difficult, I believe, by the fact that the systems that exist are the results of learning over billions of years from experiences all explicit trace of which may have vanished. This is a serious impediment as long as no fundamentally different approach is discovered for creating the same end result.

9.4 The Artificial in Artificial Intelligence

This brings us naturally to the question of what is different or artificial about AI. Artificial intelligence is not artificial because the material basis of the realization is an artifact, like a computer, as opposed to biological. That cannot be so fundamental a characteristic—computation is indifferent to the substrate on which it takes place. Also, AI is not artificial because its computational processes are different from nature. That also cannot be fundamental, since there is no impediment to emulating natural processes artificially. If there is a fundamental difference, then it must be in the way knowledge is learned from the environment. As I have argued, in nature the only way that knowledge is extracted from the world is through some kind of learning process, and in this process evolution over billions of years has had a dominating role. The difficulty of recreating the natural environments responsible for natural evolution may therefore be the most fundamental impediment to emulating natural intelligence. We may understand and be able to emulate all the algorithmic processes involved and still not be able to emulate in technology the outcome of these processes because the environments from which those algorithms learned are not reproducible.

Hence the variety of techniques currently under investigation in artificial intelligence may be viewed as attempts to replicate the outcome of the natural knowledge acquisition process by other means. In some domains these techniques will be more effective than their natural counterparts. For example, for playing chess and other games, computers conduct massive searches of game trees exploring millions of times more possible paths through a game than a human could. It is an artificial technique that happens to be more effective for chess than its natural, very different counterpart.

In Section 9.6 I will give some thoughts about how we might proceed in AI in light of what has been said. But first I will consider a different view of some central problems, and whatever insight they might hold for us.

9.5 Unsupervised Learning

An interesting task that I have not discussed so far is so-called unsupervised learning, learning where the examples come with no labels. This has been often presented symmetrically with supervised learning as a complementary phenomenon. Superficially there is something persuasive about such a complementarity. At the beginning of life on Earth surely no one was labeling examples. Consider the vision systems found in biology. They are believed to be well tuned to the natural scenes found on Earth. For example, natural objects usually have sharp boundaries and do not smoothly meld into each other. The human vision system is good at detecting these boundaries. Was this knowledge about natural scenes acquired by supervised or unsupervised means?

PAC learning as we have described it is a model of supervised learning. One of its strengths is that it is essentially assumption free. Attempts to formulate analogous theories for unsupervised learning have not been successful. In unsupervised learning the learner appears to have to make specific assumptions about what similarity means. If externally provided labels are not available, the learner has to decide which groups of objects are to be categorized as being of one kind, and which of another kind.

I hold the view that supervised learning is a powerful natural phenomenon, while unsupervised learning is not. So how do I explain all the learned knowledge in our biological circuits that cannot have been obtained by explicit labeling? My answer is evolution. Evolution in my formulation is just a form of supervised learning, where the labeling is all of one kind—fitness. To develop a vision system to distinguish friend from foe in the twilight, supervised learning was available—the target simply being vision systems of greatest utility to the evolving entity.

One can also ask about the role of unsupervised learning, not in the course of evolution, but in our brains after birth. As I have previously observed, many natural situations are self-labeled so that supervised learning applies even if there is no teacher to label the examples. A scene where we can already identify the participants as a cat and a mouse is self-labeled in that we can learn more about both cats and mice without anyone needing to label them as such. Also, our store of rules in our long-term memories imposes some metric of similarity on things we see. If we see a person on the street we may be reminded of a particular friend, not because there is some absolute metric of similarity among people, but simply because among all the circuits we happen to have, the recognition circuit for that particular friend was the one triggered. This metric is, however, ad hoc, in being totally determined by the experiences we have had and the knowledge we have acquired from them.

These arguments suggest that unsupervised learning may not be the right perspective from which to view natural phenomena. While this is the position I hold, I cannot deduce from this that unsupervised learning is a bad concept to pursue for artificial intelligence. As I have argued, there are good reasons for exploring the potential of artificial processes for knowledge acquisition that might serve in place of the natural ones.

One way we might formulate unsupervised learning for use in terms of artificial intelligence is in terms of correlation detection. We want to be able to detect which pairs of features occur together with frequencies that would be unlikely if the features were unrelated. If two particular features co-occur very frequently, then perhaps we will notice it without someone needing to point it out. If you see crowds every time you go to New York, you will probably notice this coincidence. But that is probably because the presence of crowds has a significant impact on your activities, slowing your travel or making you wait in line longer. Some other correlations we notice with greater difficulty. We handle coins all the time, but have difficulty recollecting which way the head faces, even on those we look at and use every day. In the absence of motivating factors humans often fail to notice co-occurrences even when they are glaring. Detecting correlations in general can be formulated as a computational problem, and it appears that there may be inherent computational impediments to computing correlations. If we are not good at it, then we may have a valid excuse, and if we are good then our brains are doing computations that are better than the best algorithms we can currently imagine.

The so-called light bulb problem is a formulation of the challenges inherent in detecting correlations.[8] In this problem, there is a large number N of light bulbs, each controlled by the toss of a fair coin. In each period of one second each light bulb is either on or off according to whether the coin comes up heads or tails. All the N controlling coins are independent of each other, except for just one pair. This exceptional pair is correlated so that the two light bulbs they control will be in the same state not with probability ½, as they would be if they were independent, but with some greater probability, such as ¾ or 0.51. The problem is to identify which among the $N(N-1)/2$ pairs is the correlated pair after observing the N light bulbs for a long enough period.

Detecting general correlations as formulated by the light bulb problem meets with apparent computational impediments. If one suspected that the first two light bulbs were correlated to $p = 3/4$ say, or even $p = 1/2 + \varepsilon$ (for any ε, however small, but greater than zero), then one could easily confirm or refute this by observing those two bulbs for long enough and seeing whether the fraction of times they agreed was closer to ½ or to p. It can be shown that if p is any constant greater than ½ and the light bulbs are observed for only $O(\log N)$ seconds, the pair that agree the most will, with high probability, be the correlated pair.

In general, the ith light bulb will produce an OFF/ON sequence s_i such as 00101 . . . 011. The question is whether, in order to find the correlated pair among the $N(N-1)/2$ possible pairs (i, j) of light bulbs, one has to compare every pair (s_i, s_j) of sequences and thus do order N^2 work, or whether it can be done with substantially less effort, such as $O(N^{1.5})$ or even $O(N \log N)$.

If the correlated pair is perfectly correlated, so that $p = 1$, then at every instant those two light bulbs would be always either both OFF or both ON. In that case the correlated pair can be detected in close to linear time as

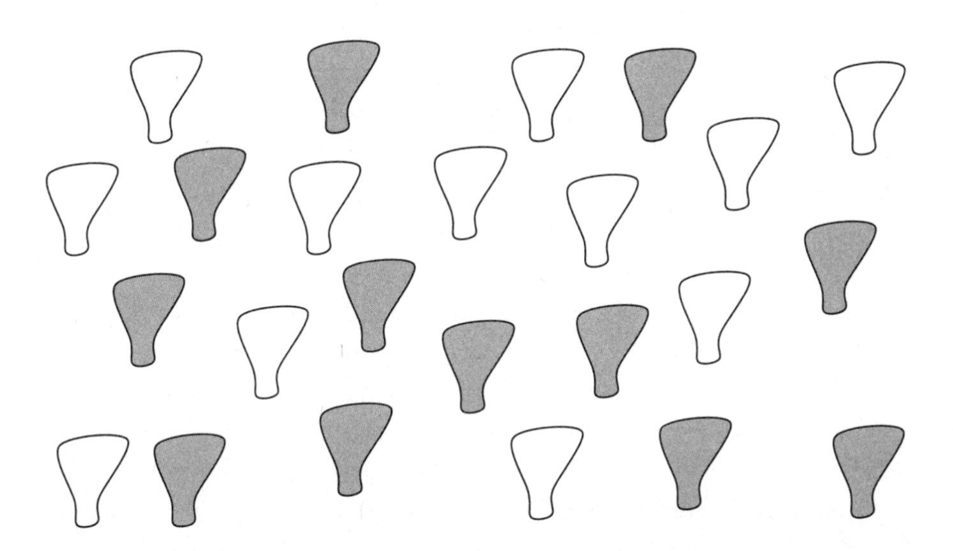

Figure 9.1 The light bulb problem is concerned with efficiently identifying which pair among a large number of flashing light bulbs is correlated. Each bulb is on half the time and off half the time, on the average. Each pair of light bulbs flash independently, except for the one correlated pair that are both on or both off with probability higher than the normal ½. The correlated pair can be identified by comparing each of the pairs of bulbs in turn. The question is whether there are methods that are more efficient than this quadratic time method.

follows. One regards the sequences s_i as binary numbers and puts them in ascending order in a list. The two perfectly correlated ones will be identical and therefore finish up next to each other in the ordered list. This ordering or sorting can be carried out in $N \log_2 N$ comparison operations on pairs of such sequences.

The computational difficulty of this correlation task, for the general case that $1/2 < p < 1$, is not too clear. Soon after the problem was initially posed, an interesting solution was found in 1989 by Ramamohan Paturi, Sanguthevar Rajasekaran, and John Reif. Their algorithm takes N^x operations, where the x decreases from 2 to 1 as p increases from ½ to 1. For $p = 0.51$, near the most challenging end of the range, this method still takes close to N^2 steps, in fact $N^{1.97}$, while there is no proof known that this task inherently needs even $N^{1.01}$ steps. Subsequent algorithms of Piotr Indyk and Rajeev Motwani, and later of Moshe Dubiner, improved on this, the latter giving $N^{1.96}$ for $p = 0.51$. Only in 2012 was a subquadratic algorithm discovered in which the exponent does not depend on p. In that year Gregory Valiant found an algorithm that works in $O(N^{1.62})$ steps for any fixed $p > 1/2$.[9]

These algorithms are somewhat complicated, advantageous only for very large numbers of light bulbs, and apparently ill suited to neural computation. It would therefore be surprising if evolution would have discovered some effective version of any of them for our brains. Hence, according to current knowledge, one would have to guess that biological systems are bad at the light bulb problem or any equivalent correlation-detection task. Humans can recognize tens of thousands of different concepts. A general capability for identifying arbitrary correlations among the millions of potential pairs of these concepts would require the brain to have a solution to the light bulb problem on a large scale. With current knowledge, it is difficult to imagine how the brain would do this.[10] A human who could demonstrate impressive light bulb detection capabilities would be executing an algorithm that we cannot currently envision.

Of course some related problems are much easier. If one only needs to detect correlated light bulbs that are physically next to each other in a line or a square array, then there is no great computational impediment. But we are asking a more general problem here. Ultimately it is conceivable, of course, that there exist near linear time algorithms for the general problem. It is equally conceivable that there are substantial inherent computational limitations that cannot be circumvented.

9.6 Artificial Intelligence—Where Next?

I repeat my belief that learning has to be at the center of the artificial intelligence enterprise. While I do not regard intelligence as a unitary phenomenon, I do believe that the problem of reasoning from learned data is a central aspect of it. Further, there are good prospects for making useful engineered systems based on combining learning and reasoning in a principled way.

I have extolled the importance of robust computational models and have claimed that for learning and evolution there is strong evidence for the existence of such models. I now emphasize that for intelligence there is no basis at present for a corresponding claim. This is consistent with thinking in other disciplines. Researchers who study intelligence tests have long recognized that the various existing tests do not all measure the same thing. Gardner has famously referred to "multiple intelligences."[11]

At a more anecdotal level, a person usually needs to be interested in what another has to say to view that other as intelligent. We find it difficult to ascribe high intelligence to people who disagree with us. This is about what one would expect if one takes robust logic as a fundamental aspect of intelligence. Intelligence is then fundamentally about how one draws conclusions from experience. If from the same set of experiences I draw a different conclusion from you, you may suspect the quality of my basic learning and reasoning algorithms. Of course, in practice, it is more likely that differences in experiences and beliefs account for our different opinions, rather than differences in the algorithms we apply.

The field of artificial intelligence traces its history back to the Turing Test, which was designed as a criterion for whether a machine could be considered to be thinking.[12] Turing's criterion of a machine passing his test was whether it could successfully impersonate a human when interacting remotely with a human. The timelessness of the Turing Test is owed, I believe, to two of its fundamental aspects. First, he proposed to have some measure of performance, by seeing how often a human subject would succeed in distinguishing the computer from what it was impersonating. Second, he was proposing a task involving unrestricted theoryless knowledge, rather than a specialized skill like chess or knowledge of chemistry.

PAC learning brings one additional idea to the table. It is that, in any particular incarnation of a Turing Test, performance is inevitably measured with respect to a particular distribution of inputs. Any particular example of the test must be conducted in a particular century in a particular place, and test for a particular task. It is not clear that if a system did pass one such instance of this test, then any more generality could be claimed for the method used, beyond success for that particular task and input distribution.

I believe that tests having the main characteristics of the Turing Test are exactly what are needed as tests for intelligent systems. But the challenge, I believe, is that of building a series of systems, for a broad variety of tasks rather than just one, all involving measurable performance on tasks that involve unconstrained theoryless knowledge. These Generalized Turing Tests would vary according to the task and distribution.

The formulation I gave of robust logic is along these lines. It encompasses the two aspects of the Turing Test that I highlighted—of having some measure by which to gauge performance, namely the accuracy of predictions, and of being applicable to general theoryless knowledge. It gives a concrete framework for reasoning in the common sense context where knowledge is theoryless and learned. The behavior of any system built along these principles will depend on the environment in which it is trained. In that sense intelligent behavior is indeed relative, and in the eye of the beholder. Of course, one would hope that the functionality of the basic learning and reasoning algorithms is more absolute. That would imply that there is something theoryful in intelligence.

Turing discussed the dilemma between having machines programmed and having them learn. He recognized the importance of the latter.[13] After his death, however, the argument gained ascendency that a machine that learned will, after it has learned, be executing some fixed program, and therefore that learned program could have been equally created by a programmer. This argument is best regarded as spurious. In most instances where machine learning algorithms are already used successfully, it would not be practicable to replace the final product by one explicitly programmed. This is because there is no alternative way known for obtaining the program learned, with all the associated parameter values, other than by a learning process.

Why exactly are the results of human learning so difficult to replicate otherwise? I believe that there is essentially one reason, one already mentioned. When explicitly programming a machine or teaching a person, one is not starting from a blank slate, but building on features for which the machine or human already has programs. To be able to write an explicit program a programmer needs to understand exactly what the features already realized in the system do. Otherwise there is the risk of creating undesirable new consequences or side effects.

Learning has the important advantage that the process operates relative to whatever functions are already being realized in the system. Each example presented by an instructor in a course will, for each student, and for each concept that that student already recognizes, either be recognized as an example of that concept, or not. The example will be related to the student's previous knowledge in this way, even if the teacher does not know exactly what that relationship is. In this sense teaching is a much more robust activity than programming, since the learner's previous knowledge will be applied automatically to the interaction. A programmer has to go much further in adapting to the system being programmed than a teacher has to the learner. This is particularly important when the only available teacher is a passive environment.

An advantage of learning therefore is that it interfaces directly between the possibly complex current state of knowledge of the learner and the invariably complex outside world. Learning can be accomplished without needing anyone to explicitly understand either the state of the learner or the complexities of the world. A programmer would need to understand both to be successful.

Taking the centrality of learning now as given, I believe that the task of constructing intelligent systems can be divided into three parts. The first consists of providing generic learning and reasoning algorithms along the lines we have described in previous chapters. The second is some architecture that describes how to use these algorithms in combination, for example to manipulate a mind's eye. The third part is that of producing appropriate teaching materials, the examples from which to learn.

This last issue is rarely discussed in artificial intelligence, though it dominates in human education. Educational institutions invest much effort on curricula, in selecting and organizing the material to be taught. A random presentation of relevant material is not considered adequate. It is only reasonable to expect that once artificial intelligence becomes more learning centered, as no doubt one day it will be, it will become obvious that a computer's curriculum should be prepared with no less care than a child's.

9.7 Need We Fear Artificial Intelligence?

There may be some good news for humans in the fact that one can be intelligent in many different ways. It gives us hope that we may endow robots with intelligence superior to ours but only in directions that are useful and not threatening to us. Also, it makes it clear that there is no good reason to want to make robots that are exactly like humans.

The most singular capability of living organisms on Earth must be that of survival. Anything that survives for billions of years, and many millions of generations, must be good at it. Fortunately, there is no reason for us to endow robots with this same capability. Even if their intelligence becomes superior to ours in a wide range of measures, there is no reason to believe that they would deploy this in the interests of their survival over ours unless we go out of our way to make them do just that. We have limited fear of domesticated animals. We do not necessarily have to fear intelligent robots either. They will not resist being switched off, unless we provide them with the same heritage of extreme survival training that our own ancestors had been subject to on Earth.

Chapter Ten

Questions

Computers are boring. They only give answers.
Pablo Picasso

10.1 Science

It was a spring day. I was a mathematics student in Cambridge in England. As on most mornings, I went to the Arts School where lectures were held. Instruction was well organized though a little predictable. One could tell what topic each lecture was on from whether the day of the week was even or odd, and whether the hour was even or odd. Otherwise these mornings were largely uneventful.

That day, however, was a little different. An older man whom I did not recognize was standing at the front. As we soon found out, it was Paul Dirac about to give his retirement lecture. Even undergraduates in the audience like me would have known that Dirac had discovered a significant fraction of the theory of quantum mechanics, and hence was a major figure in twentieth-century physics. They would also have known that he had written a revered textbook, was occupying the Lucasian chair once held by Newton, and perhaps also that he had received a Nobel Prize many decades earlier. However, anyone expecting any element of satisfaction or triumph in this retirement lecture was to be disappointed. His theme was regret.

He explained how in his work on quantum mechanics whenever he discovered something he published it and waited for the discovery to receive some acceptance from others. This validation would be in the form of others following up his work to make further discoveries. His regret was that he was not able to take his own work seriously enough that he could build on it and take the next steps himself without waiting for others to do so. Had he not had this psychological impediment, he would have been able to achieve more.

I have always believed that science is totally impersonal. If it matters who discovered it, or who said it, or how it is said, then it is not science. However, as I learned that day, the pursuit of science can be a highly personal matter. I have wondered ever since how such personal pursuits can come together to build the overall edifice that science is.

10.2 A More Strongly Ecorithmic Future

Much has changed since we humans first arrived on the scene. We now grow food communally and cook it. Many of us live in highly organized communities of millions of individuals, with water and electric power piped into our homes, and waste piped out. Most recently, we have also had information piped both in and out at alarming speed.

Until the emergence of humans there was an equilibrium on Earth between Darwinian evolution, on the one hand, and the more general learning algorithms of individual organisms, on the other. Life forms evolved. The result of evolution was to give living organisms certain learning capabilities that enabled them as individuals to adapt to the world during their life. Some material, such as birdsong, was handed down, with each generation learning from the previous one. However, Darwinian evolution remained the mechanism responsible for essentially all significant new adaptations that outlasted the lives of individuals. The learning and reasoning carried out by an organism during its life had limited impact that outlived the individual.

With human civilization this equilibrium tilted significantly toward more general ecorithms and away from the limited ones that characterize Darwinian evolution. Being the more powerful, these general ecorithms hastened the pace of change. Individuals already could teach their children and pass on knowledge that they had learned, but with the advent of civilization, they could create songs, poems, paintings, and books capable of outlasting the lives of their individual creators. They could be used widely and indefinitely as teaching materials for later generations. In this way the knowledge inherent in the intellectual work of individuals became a shared resource and part of culture. An individual could for the first time acquire circuits that represented the knowledge gained from experience by thousands of different people, without having to go through their experiences. One can regard the massive accumulation of collective knowledge that this made possible through many generations as learnable target pursuit on a scale that had not been previously possible when it had been limited to single individuals learning.

Culture also undergoes change or evolution, but this change is no longer limited by Darwinian principles. Among some remarkable accompaniments of this cultural evolution have been the discovery and exploitation of science, and the resulting changes in how we live.

The breeding of plants and animals of greater and greater value to humans has been important for thousands of years, and it now enables a far larger human population to be maintained than would otherwise be possible. As Darwin emphasized, such artificial selection is essentially the same process as natural selection, except the choice of fitness is made artificially. In our terms, they are both subject to the same evolutionary ecorithms, except that the former is driven by a teacher or environment that redefines the target at will.

It is one thing to tap the power of ecorithms in such an inadvertent fashion. Now that we better understand their potential and might more fully exploit it through technology, ecorithms stand to reshape our civilization more dramatically than they have heretofore. A small hint of that power can be gleaned from our current experience with Web search engines. The least educated person today has immeasurably more knowledge on tap than the most educated had a mere twenty years ago. The basic algorithms used in search engines, of sifting out Web pages that contain certain words and presenting those first that are most likely to be of interest, are no more than algorithms. The awesome power of the Web is due entirely to the scale on which it operates—billions of Web pages. The same algorithms run on a small, specialized dataset would arouse little fascination.

Our intuitions about enormous datasets, and the information that may be extracted from them, are still in their infancy. But there is little doubt that the existence of such datasets, and the possibility of mining them by learning algorithms, is a major new facet of our world. Once more sophisticated learning algorithms are let loose on all the information available in the world, we may glean knowledge that is not obtainable any other way. There seems little doubt, for example, that if detailed medical histories and personal habits on billions of people were to become available in consistent electronic form, then significant new discoveries regarding medicine and health care could be made. But the opportunity seems greater still. As, aided by computer technology, we use ecorithms in artificial systems to their fuller potential, we can hope for and pursue results as astonishing as the previous two phases of evolution, the biological and the cultural, have themselves produced.

10.3 How to Act?

When we talk and write, we are reporting on the output of our brain circuits. These circuits generally do not compute anything theoryful or preordained. They are the results of learning processes prompted by events over billions of years and tuned to perform some theoryless task. The exact nature of the input/output functions that these circuits realize is beyond our current understanding.

Nevertheless, there are some generalizations we can make about these circuits. In the area of science, for example, the outputs of the brain circuits of experts are usually in high agreement with each other. Also, they often predict with great accuracy the outcomes of future experiments. Technological products based on these predictions generally work as expected. In areas of such high agreement and high predictive accuracy, it is natural to take seriously the outputs of these brain circuits as having some reality beyond being just expressions of personal feelings. One is certainly ill advised to bet against them given the overwhelming empirical evidence of their accuracy.

More broadly we have to take seriously the utterances of experts about situations in which they have made accurate predictions in the past. As we know, our plumber, doctor, and car mechanic should be taken as seriously as their respective predictive records warrant. Even when no such records are available, we may be well advised to follow the advice of those who have experience. Predictions that fit past data even for short periods and only in one place may be the best kind of advice that we can get, whether or not those predictions have a theoryful basis. In acknowledging this, we are accepting that the PAC sense of learning is the best that we can generally expect to achieve given that the world is as complex and uncertain as it is.

Such generalizations only take us so far, however, and they tell us very little about what we should make of the many areas of human concern where no consensus opinion is in sight and where there is little evidence that anyone's brain circuits are highly predictive. We do not need any reminders of the lack of permanent standing of the many beliefs that have been entertained over the centuries. Attitudes to politics, race, gender, religion, art, economics, and health have gyrated wildly, and few would, or even could, subscribe to more than a small fraction of the many contradictory opinions that are currently being maintained around the world. We can always take an individual's pronouncement on a question in such areas as an expression of personal feelings and give it the respect that it deserves on that basis. However, we need to be cautious. We should not adapt our views too eagerly to conform to that of others. We know that our circuits can rush to judgment on scant evidence. We must remember that theirs can do the same.

In the opposite direction, we should also be cautious in taking on the responsibility of influencing others. It is difficult to imagine what justification we could have for exerting influence if no validity can be ascribed to our views beyond that of personal feelings. So, too, would it do well for political or policy-oriented organizations, whose main goal is to promote particular personal opinions, to exercise restraint, especially as we live in an age when information can be disseminated cheaply on an unprecedented scale. As Churchill put it, "The empires of the future are the empires of the mind."[1] Can society permit any organization that promotes the theoryless to grow to have imperial influence?

10.4 Mysteries

An important aspect of human personal experience is that of the mysterious—of things we do not understand and know we do not. Scientific theories have mysteries too, namely those aspects that are fundamental to the theories but not explained by them. Such mysteries, more often than not, are known to the theory's originators. One celebrated example is the identity of inertial and gravitational mass at the heart of Newtonian physics. Newton had equated the former, the resistance of a body to being accelerated by a force, to the latter, the gravitational pull exerted on it by another object. This seeming coincidence is a central part of his theory of mechanics, but one for which he knew he had no explanation. Not until Einstein's general theory of relativity was this removed from the realm of the mysterious. Earlier I quoted Eugene Wigner's comments on the effectiveness of mathematics in the physical sciences. Even as we have exploited mathematics to gain an ever more accurately predictive understanding of the physical world, the question of why this approach works—of why abstract mathematical thought should illuminate the physical world—has, if anything, become a greater and greater mystery.

I have argued that ecorithms offer an avenue toward a greater understanding of the phenomena of evolution and cognition, but ecorithms have their mysteries too. In concluding, it is fitting to be more explicit about some of these.

First, the whole theory is based on the notion that there are mathematical theories of computation that explain the limits of computing, learning, and evolving. This is in turn based on the separate notion that there are robust models of computation that capture these phenomena. There is much evidence for both of these notions being true. However, as with analogous statements in the physical sciences, why these notions should hold is totally mysterious, and there is no indication that we have a viable approach toward understanding them.

Second, though I have framed both evolution and cognition as manifestations of ecorithms, it is currently unknown what the major ecorithms that operate in biology are. These particular current mysteries I regard as potentially more tractable than the first set. I am confident that, as the inertial versus gravitational issue was eventually resolved, these will be also in the course of future enquiry. All the sciences will need to contribute.

One may ask why human progress has been so much more successful in science than in other areas. In my view credit goes entirely to the nature of science itself rather than to anything special about the humans who pursue it. Science does make progress, and this ultimate outcome is not endangered by any vagaries in the process. Science is supported by robust scaffolding that we can ascend one step at a time. While each step holds surprises and may be difficult to take, it is the overall unity found when looking back that is breathtaking.

The science of learning mechanisms explores how a computationally limited entity can succeed in a world that is too complex for it to model. It focuses on three things: computationally bounded entities, successful action in a complex world, and, most importantly, the relationship between the two.

Darwin observed that there is grandeur in the view that life is the result of a simple evolutionary mechanism. The ecorithmic formulation aims to understand more explicitly the mechanisms of evolution, as well as of its child, cognition. Further progress here will surely enhance rather than diminish this view of grandeur. It will contribute to bringing an account of our intimate natures within the scope of science, and to explaining why such an account can be discovered by humble creatures of its laws.

Notes

Chapter 1

1. This account is adapted from Fritz Alt, "Archeology of Computers: Reminiscences, 1945–47," *Communications of the ACM* 15, no. 7 (July 1972): 693–694.

2. A. M. Turing, "On Computable Numbers, with an Application to the *Entscheidungsproblem*," *Proceedings of the London Mathematical Society*, Ser. 2, 42 (1936–1937): 230–265.

3. The concept of PAC learning was introduced in L. G. Valiant, "A Theory of the Learnable," *Communications of the ACM* 27, no. 11 (1984): 1134–1142. The concept was subsequently named "probably approximately correct" in D. Angluin and P. Laird, "Learning from Noisy Examples," *Machine Learning* 2 (1987): 343–370.

4. N. Taleb, *The Black Swan* (New York: Random House, 2007); D. Kahneman, *Thinking Fast and Slow* (New York: Farrar, Straus and Giroux, 2011).

Chapter 2

1. In "How U.N. Chief Discovered U.S., and Earmuffs," *New York Times* interview, January 7, 1997.

2. From Foreword by Donald E. Knuth to M. Petkovšek, H. Wilf, and D. Zeilberger, *A=B* (Wellesley, MA: A.K. Peters, 1997).

3. Alfred Russel Wallace proposed the same theory independently in "On the Tendency of Species to Form Varieties, and on the Perpetuation of Varieties and Species by Natural Means of Selection," *Journal of the Proceedings of the Linnean Society of London, Zoology* 3 (1858): 53–62. Because of Darwin's much more detailed exposition in his *On the Origin of Species* (London: Murray, 1959), the theory has become more closely identified with his name.

4. The Weald is an area stretching from Hampshire in the west to Kent in the east, between the North and South Downs in southern England.

5. J. Marchant, *Alfred Russel Wallace, Letters and Reminiscences*, vol. I (London: Cassell, 1916), 242, letter dated April 14, 1869.

6. Lord Kelvin (William Thomson), "The Age of the Earth as an Abode Fitted for Life," *Journal of the Transactions of the Victoria Institute* 31 (1899): 11–35.

Chapter 3

1. The historical context and the related work of contemporaries Gödel, Post, Church, and others are described in M. Davis (ed.), *The Undecidable: Basic Papers on Undecidable Propositions, Unsolvable Problems and Computable Functions* (Mineola, NY: Dover, 2004).

2. To be more precise, Turing's paper refers to problems equivalent to the Halting Problem, including the Printing Problem, which asks whether a certain symbol will be ever written.

3. K. Gödel, "Remarks Before the Princeton Bicentennial Conference on Problems in Mathematics" (1946), reprinted in Davis (ed.), *The Undecidable*, 84–88.

4. Eugene Wigner, "The Unreasonable Effectiveness of Mathematics in the Natural Sciences," in *Communications in Pure and Applied Mathematics*, vol. 13, no. 1 (February 1960). New York: John Wiley & Sons.

5. The term "computational complexity" was coined by Juris Hartmanis and Richard Stearns in their pioneering study of the time and space requirements of Turing machine computations. Earlier, in 1960, Michael Rabin had given an axiomatic theory of this phenomenon. An earlier reference still, in the context of cryptography, is a letter from John Nash to the National Security Agency in 1955 (www.nsa.gov/public_info/press_room/2012/nash_exhibit.shtml). Comprehensive expositions of this field can be found in C. H. Papadimitriou, *Computational Complexity* (Boston: Addison-Wesley, 1994); O. Goldreich, *Computational Complexity: A Conceptual Perspective* (New York: Cambridge University Press, 2008); and S. Arora and B. Barak, *Complexity Theory: A Modern Approach* (New York: Cambridge University Press, 2009).

6. A function $f(n)$ is $O(g(n))$ if for some constant k and for all $n > 0, f(n) < kg(n)$. If one changes the basis of the number representation from 10 to another number, such as 2 for the case of binary arithmetic that computers use, the long multiplication algorithm is still $O(n^2)$ steps.

7. Strictly speaking, P is usually defined only for problems with yes/no answers. For simplicity, in this book we will also use it to include problems with many bit outputs, such as integer multiplication, if computing each bit of the output is a P problem in the more standard sense, and there are only polynomially many output bits.

8. A. Karatsuba and Yu. Ofman, "Multiplication of Multi-Digit Numbers on Automata," *Soviet Physics Doklady* 7 (1963): 595–596.

9. A. Schönhage and V. Strassen, "Schnelle Multiplikation grosser Zahlen," *Computing* 7 (1971): 281–292. The runtime of their algorithm is $O(n \log n \text{ loglog } n)$ steps, an expression that grows more slowly than $n^{1.001}$, or $n^{1+\varepsilon}$ for any positive ε. In 2007 this was slightly improved by Martin Fürer to a function that still grows a little more slowly than $n \log n$.

10. For polynomial time algorithms for testing primality, see Robert Solovay and Volker Strassen, "A Fast Monte-Carlo Test for Primality," *SIAM Journal on Computing* 6, no. 1 (1977): 84–85; Gary L. Miller, "Riemann's Hypothesis and Tests for Primality," *Journal of Computer and System Sciences* 13, no. 3 (1976): 300–317;

M. O. Rabin, "Probabilistic Algorithm for Testing Primality," *Journal of Number Theory* 12, no. 1 (1980): 128–138. A deterministic algorithm with higher but still polynomial complexity was found more recently: M. Agrawal, N. Kayal, and N. Saxena, "PRIMES Is in P," *Annals of Mathematics* 160, no. 2 (2004): 781–793.

11. R. Rivest, A. Shamir, and L. Adleman, "A Method for Obtaining Digital Signatures and Public-Key Cryptosystems," *Communications of the ACM* 21, no. 2 (1978): 120–126. A general approach to relating cryptography and complexity theory is given in S. Goldwasser and S. Micali, "Probabilistic Encryption," *Journal of Computer and System Sciences* 28, no. 2 (1984): 270–299.

12. Turing had used the phrase "intellectual search" for a seemingly similar concept, but without an explicit polynomial criterion: A. M. Turing, "Intelligent Machinery" (unpublished manuscript, 1948), reproduced in B. J. Copeland, *The Essential Turing* (Oxford: Oxford University Press, 2004), 410–432.

13. The notion of NP-completeness was introduced in S. A. Cook, "The Complexity of Theorem Proving Procedures," *Proceedings, Third Annual ACM Symposium on the Theory of Computing* (1971): 151–158. The range of NP-complete problems was greatly extended by R. M. Karp, "Reducibility among Combinatorial Problems," in Raymond E. Miller and James W. Thatcher (eds.), *Complexity of Computer Computations* (New York: Plenum Press, 1972), 85–103. A parallel development occurred in the Soviet Union: L. Levin, "Universal Search Problems," *Problems of Information Transmission* 9, no. 3 (1973): 265–266 (in Russian), translated into English by B. A. Trakhtenbrot, "A Survey of Russian Approaches to Perebor (Brute-Force Searches) Algorithms," *Annals of the History of Computing* 6, no. 4 (1984): 384–400. The NP-completeness phenomenon as seen a few years later is excellently described in M. R. Garey and D. S. Johnson, *Computers and Intractability: A Guide to the Theory of NP-Completeness* (New York: W.H. Freeman, 1979).

14. K. L. Manders and L. M. Adleman, "NP-Complete Decision Problems for Quadratic Polynomials," *Proceedings, Eighth Annual ACM Symposium on the Theory of Computing* (1976): 23–29.

15. L. G. Valiant, "The Complexity of Computing the Permanent," *Theoretical Computer Science* 8 (1979): 189–201; L. G. Valiant, "The Complexity of Enumeration and Reliability Problems," *SIAM Journal on Computing* 8, no. 3 (1979): 410–421.

16. E. Bernstein and U. Vazirani, "Quantum Complexity Theory," *SIAM Journal on Computing* 26, no. 5 (1997): 1411–1473. This paper introduced the quantum class BQP. Earlier formulations of quantum computation had been given by R. P. Feynman, "Simulating Physics with Computers," *International Journal of Theoretical Physics* 21 (1982): 467–488, and D. Deutsch, "Quantum Theory, the Church-Turing Principle and the Universal Quantum Computer," *Proceedings of the Royal Society of London, Series A, Mathematical and Physical Sciences* 400 (1985): 97–117.

17. Note that each of these classes contains functions with only yes/no values, except for #P, which produce numbers. The PAC class is illustrated as a subclass of

P, but one could extend it into BQP, for example. It is currently unknown, for any pair of the classes illustrated, whether they are of equal extent to within polynomial time deterministic reductions. Every two of the classes shown is widely conjectured to be different, except for the P =? BPP question, for which some suggestion of the contrary conjecture can be found in R. Impagliazzo and A. Wigderson, "P = BPP if E Requires Exponential Circuits: Derandomizing the XOR Lemma," *Proceedings of the 29th ACM Symposium on Theory of Computing* (1997): 220–229. Of course, taking a position on any these mathematical conjectures is a theoryless activity, and this author is not doing that here.

18. F. Rosenblatt, *Principles of Neurodynamics: Perceptrons and the Theory of Brain Mechanisms* (Washington, DC: Spartan Books, 1962). A detailed analysis is given by M. Minsky and S. Papert, *Perceptrons: An Introduction to Computational Geometry*, 2nd ed. (Cambridge, MA: MIT Press, 1972).

19. This example is suggested by a dataset on iris varieties from R. A. Fisher, "The Use of Multiple Measurements in Taxonomic Problems," *Annual Eugenics* 7, part II (1936): 179–188.

20. Without loss of generality we can make the right-hand side of any perceptron 0 by adding an extra variable to the left-hand side and extending each example to have the fixed value 1 for this last variable. Figure 3.7 implements this idea to find the separator $3x - 6y > 1$ for the six points listed in the rubric of Figure 3.6. Note that the inequality $2x - 3y > 2$ illustrated there also satisfies these six examples.

Chapter 4

1. Eddington made this remark in Leicester, UK, at the annual meeting of the British Association for the Advancement of Science: "Star Birth Sudden Lemaître Asserts," *New York Times*, September 12, 1933.

2. A. M. Turing, "The Chemical Basis of Morphogenesis," *Philosophical Transactions of the Royal Society of London. Series B, Biological Sciences* 237, no. 641 (August 1952): 37–72.

3. This viewpoint also has received support from within the biological sciences community: P. Nurse, "Life, Logic and Information," *Nature* 454 (2008): 424–426.

4. M. H. A. Newman, "Alan Mathison Turing, 1912–1954," *Biographical Memoirs of Fellows of the Royal Society* 1 (1955): 253–263.

Chapter 5

1. A. M. Turing, "Solvable and Unsolvable Problems," *Science News* 31 (1954): 7–23.

2. The nature of the computations that brain-like systems are capable of executing within realistic resource limitations deserves separate investigation: L. G. Valiant, *Circuits of the Mind* (New York: Oxford University Press, 1994, 2000); L. G. Valiant, "Memorization and Association on a Realistic Neural Model," *Neural Computation*

17, no. 3 (2005): 527–555. Various failings of human memory from an experimental psychology perspective are described in D. Schacter, *The Seven Sins of Memory: How the Mind Forgets and Remembers* (New York: Houghton Mifflin, 2002).

3. Aristotle, *Posterior Analytics, Book I*, translated by G. R. G. Mure (eBooks@ Adelaide, 2007).

4. P. Hallie (ed.), *Selections from the Major Writings on Skepticism, Man and God*, translated by S. Etheridge (Indianapolis, IN: Hackett, 1985), 105.

5. A calculation shows that a sample size $(2/error) \times (n + \log_e(1/error))$ suffices: L. G. Valiant, "A Theory of the Learnable," *Communications of the ACM* 27, no. 11 (1984): 1134–1142.

6. A sample size similar in terms of *n* and *error* to that in Note 5, above, still suffices.

7. The study of elimination, but without any quantitative analysis of what it achieves, has a long history: John Stuart Mill, *A System of Logic* (London: John W. Parker, 1843).

8. A purely computational theory is given by E. M. Gold, "Language Identification in the Limit," *Information and Control* 10 (1967): 447–474. A statistical theory is provided in V. N. Vapnik, *The Nature of Statistical Learning Theory* (New York: Springer-Verlag, 2000), and T. Hastie, R. Tibshirani, and J. H. Friedman, *The Elements of Statistical Learning* (New York: Springer-Verlag, 2001).

9. More details on PAC learning and its extensions can be found in M. J. Kearns and U. Vazirani, *An Introduction to Computational Learning Theory* (Cambridge, MA: MIT Press, 1994).

10. The Occam formulation is from A. Blumer, A. Ehrenfeucht, D. Haussler, and M. K. Warmuth, "Occam's Razor," *Information Processing Letters* 24 (1987): 377–380. It exemplifies how the purely statistical criterion of learnability is almost tautological if examples and hypotheses are to be represented discretely, which in reality they always are. For infinite representations, such as real numbers, an analogous treatment is still possible, but more involved, via, for example, the VC dimension: V. Vapnik and A. Chervonenkis, "On the Uniform Convergence of Relative Frequencies of Events to Their Probabilities," *Theory of Probability and Its Applications* 16, no. 2 (1971): 264–280; A. Blumer, A. Ehrenfeucht, D. Haussler, and M. K. Warmuth, "Learnability and the Vapnik–Chervonenkis Dimension," *Journal of the ACM* 36, no. 4 (1989): 929–965. Earlier work using related concepts: Thomas M. Cover, "Capacity Problems for Linear Machines," in L. Kanal (ed.), *Pattern Recognition* (Washington, DC: Thompson Book Co., 1968).

11. Such a general lower bound on the number of examples needed for learning is given in A. Ehrenfeucht, D. Haussler, M. Kearns, and L. G. Valiant, "A General Lower Bound on the Number of Examples Needed for Learning," *Information and Computation* 82, no. 2 (1989): 247–261.

12. The first publication of the notion of public-key cryptosystems was W. Diffie and M. E. Hellman, "New Directions in Cryptography," *IEEE Transactions on Information Theory* IT-22 (November 1976): 644–654. The RSA system is from

R. Rivest, A. Shamir, and L. Adleman, "A Method for Obtaining Digital Signatures and Public-Key Cryptosystems," *Communications of the ACM* 21, no. 2 (1978): 120–126. There had been earlier unpublished work on these concepts by James Ellis, Clifford Cocks, and Malcolm Williamson at the Government Communications Headquarters in the UK, and also by Ralph Merkle at UC Berkeley.

13. Here we are regarding the decryption function as outputting a set of yes/no functions, namely the bits of the original message, and each one would be learned. In any public-key cryptosystem the encryption algorithm is available to all.

14. N. Chomsky, "Three Models for the Description of Language," *IRE Transactions on Information Theory* 2 (1956): 113–124.

15. M. Kearns and L. G. Valiant, "Cryptographic Limitations on Learning Boolean Formulae and Finite Automata," *Journal of the ACM* 41, no. 1 (1994): 67–95. Preliminary version in *Proceedings of the 21st ACM Symposium on Theory of Computing* (1989): 433–444.

16. A. Klivans and R. Servedio, "Learning DNF in time $2^{\tilde{O}(n^{1/3})}$," *Journal of Computer and System Sciences* 68, no. 2 (2004): 303–318.

17. An elegant attribute-efficient algorithm, called Winnow, for learning disjunctions is given in N. Littlestone, "Learning Quickly When Irrelevant Attributes Abound: A New Linear-Threshold Algorithm," *Machine Learning* 2, no. 4 (1988): 285–318. It resembles the perceptron algorithm, but the weights are updated by multiplying rather than by adding appropriate quantities. See also Avrim Blum, "Learning Boolean Functions in an Infinite Attribute Space," *Machine Learning* 9 (1992): 373–386.

18. R. I. Arriaga and S. Vempala, "An Algorithmic Theory of Learning: Robust Concepts and Random Projection," *Proceedings of the 40th IEEE Symposium on Foundations of Computer Science (FOCS)* (1999): 616–623.

Chapter 6

1. The Royal Tyrrell Museum in Drumheller, Alberta, Canada, is instructive.

2. Questions about the absence of quantitative explanations in evolutionary theory were raised by various authors in P. S. Moorhead and M. M. Kaplan (eds.), *Mathematical Challenges to the Neo-Darwinian Interpretation of Evolution: A Symposium, Philadelphia, April 1966* (Philadelphia: Wistar Institute Press, 1967). An attempt to address the issue in the context of the eye is given in D. E. Nilsson and S. Pelger, "A Pessimistic Estimate of the Time Required for an Eye to Evolve," *Proceedings: Biological Sciences* 256 (1994): 53–58.

3. R. A. Fisher, *The Genetical Theory of Natural Selection* (Oxford: Oxford University Press, 1930); S. Wright, *Evolution and the Genetics of Populations, A Treatise* (Chicago: University of Chicago Press, 1968–1978).

4. The award winners in an annual competition for genetic programming can be found at http://www.genetic-programming.org/combined.html.

5. The term comes from the title of Julian Huxley's book *Evolution: The Modern Synthesis* (1942). It refers to the synthesis reached in the preceding decade of three disparate strands: natural selection as understood by Darwin and Wallace, Mendelian genetics, and the population biology of R. A. Fisher, J. B. S. Haldane, and Sewall Wright.

6. U. Alon, *An Introduction to Systems Biology* (Boca Raton, FL: CRC Press, 2006).

7. For simplicity, our terminology assumes deterministic input functions. However, the discussion is the same if we allow randomized or even quantum transitions.

8. C. D. Allis, T. Jenuwein, and D. Reinberg, *Epigenetics* (Cold Spring Harbor, NY: CSHL Press, 2007).

9. The possible impact on natural selection of learning during life has been explored informally since the nineteenth century: M. J. Baldwin, "A New Factor in Evolution," *The American Naturalist* 30, no. 354 (June 1896): 441–451; G. E. Hinton and S. J. Nowlan, "How Learning Can Guide Evolution," *Complex Systems* 1 (1987): 495–502.

10. N. Eldredge and S. J. Gould, "Punctuated Equilibria: An Alternative to Phyletic Gradualism," in T. J. M. Schopf (ed.), *Models in Paleobiology* (San Francisco: Freeman, Cooper and Company, 1972), 82–115.

11. A. R. Wallace, "The Measurement of Geological Time," *Nature*, 17 (1870): 399–341, 452–455.

12. We are oversimplifying in this exposition, as we did also with regard to learning in Chapter 5, in not distinguishing the class C of target ideal functions from the representations used for the hypotheses. The latter may be different from the former.

13. The evolvability model described here is from L. G. Valiant, "Evolvability," *Journal of the ACM* 56, no. 1 (2009): 3:1–3:21. (Earlier versions: *Proceedings of the 32nd International Symposium on Mathematical Foundations of Computer Science,* August 26–31, 2007; Český Krumlov, Czech Republic, *Lecture Notes in Computer Science,* vol. 4708 [New York: Springer-Verlag, 2007], 22–43; and *Electronic Colloquium on Computational Complexity,* Report 120, September 2006.)

14. M. J. Kearns, "Efficient Noise-Tolerant Learning from Statistical Queries," *Journal of the ACM* 45, no. 6 (1998): 983–1006.

15. V. Feldman, "Distribution-Independent Evolvability of Linear Threshold Functions," *Journal of Machine Learning Research—Proceedings Track* 19 (2011): 253–272.

16. M. J. Kearns, "Efficient Noise-Tolerant Learning from Statistical Queries."

17. V. Feldman, "Robustness of Evolvability," *Proceedings of the 22nd Annual Conference on Learning Theory, Montreal, Quebec, Canada* (2009). See also V. Feldman, "Evolvability from Learning Algorithms," *Proceedings of the 40th Annual ACM Symposium on Theory of Computing* (2008): 619–628.

18. L. Michael, "Evolvability via the Fourier Transform" (manuscript, 2007). Also *Theoretical Computer Science* 462 (2012): 88–98.

19. V. Feldman, "A Complete Characterization of Statistical Query Learning with Applications to Evolvability," *50th Annual IEEE Symposium on Foundations of Computer Science* (2009): 375–384.

20. P. Valiant, "Distribution Free Evolvability of Polynomial Functions over All Convex Loss Functions," *Proceedings of the 3rd Symposium on Innovations in Theoretical Computer Science* (2012): 142–148.

21. Ibid.

22. V. Kanade, "Evolution with Recombination," *52nd Annual IEEE Symposium on Foundations of Computer Science* (2011): 837–846.

23. R. A. Fisher, *The Genetical Theory of Natural Selection* (Oxford: Clarendon Press, 1930); J. Maynard Smith, *The Evolution of Sex* (Cambridge: Cambridge University Press, 1978); A. Livnat, C. H. Papadimitriou, J. Dushoff, and M. W. Feldman, "A Mixability Theory of the Role of Sex in Evolution" *PNAS* 105, no. 50 (2008): 19803–19808.

24. Some evolution algorithms can be shown to be able to tolerate a slowly changing world: V. Kanade, L. G. Valiant, and J. Wortman Vaughan, "Evolution with Drifting Targets," *Conference on Learning Theory* (2010): 155–167.

Chapter 7

1. I. Ayres, *Super Crunchers: Why Thinking-by-Numbers Is the New Way to Be Smart* (New York: Bantam, 2007).

2. This formulation is further described in L. G. Valiant, *Circuits of the Mind* (New York: Oxford University Press, 1994, 2000).

3. D. B. Lenat, "CYC: A Large-Scale Investment in Knowledge Infrastructure," *Communications of the ACM* 38, no. 11 (1995): 32–38.

4. Bayesian models and inference are described in J. Pearl, *Probabilistic Reasoning in Intelligent Systems* (San Francisco: Morgan Kaufmann Publishers, 1988). Further studies of reasoning in uncertain contexts can be found in the *Uncertainty in Artificial Intelligence* conference series. It has been found empirically that in applications where large amounts of general knowledge need to be modeled, some learning component is essential, as, for example, in IBM's Watson system for the *Jeopardy!* contest.

5. D. Angluin and P. Laird, "Learning from Noisy Examples," *Machine Learning* 2 (1987): 343–370. A generic approach to making learning algorithms resistant to one kind of noise is given in Michael J. Kearns, "Efficient Noise-Tolerant Learning from Statistical Queries," *Journal of the ACM* 45, no. 6 (1998): 983–1006.

6. L. G. Valiant, "Robust Logics," *Artificial Intelligence Journal* 117 (2000): 231–253.

7. G. A. Miller, "The Magical Number Seven Plus or Minus Two," *The Psychological Review* 63 (1956): 81–97.

8. F. Galton, *Inquiries into Human Faculty and Its Development*, 1st ed. (London: Macmillan, 1883).

9. I. Biederman, "Recognition–by-Components: A Theory of Human Image Understanding," *Psychological Review* 94 (1987): 115–147.

10. In this work, discussions of the brain are in terms of what it needs to do, and not how it does it. For the latter, see Note 2 to Chapter 5.

11. Reaction time experiments show that the visual system can recognize what object is in a scene extremely rapidly: S. J. Thorpe, D. Fize, and C. Marlot, "Speed of Processing in the Human Visual System," *Nature* 381 (1996): 520–522.

12. N. Littlestone, "Learning Quickly When Irrelevant Attributes Abound: A New Linear-Threshold Algorithm," *Machine Learning* 2, no. 4 (1988): 285–318.

13. The one constraint needed for the reasoning part to be polynomial in terms of the relevant parameters (such as the size of the rules) is that the number of arguments in all the relations be bounded by a constant, since the reasoning process is exponential in that quantity.

14. A fuller discussion of how robust logic might be used in intelligent systems is given in L. G. Valiant, "Knowledge Infusion," *Proceedings of the 21st National Conference on Artificial Intelligence*, July 16–20, Boston, MA (Menlo Park, CA: AAAI Press, 2006), 1546–1551. Some experimental results are reported in L. Michael and L. G. Valiant, "A First Experimental Demonstration of Massive Knowledge Infusion, *Proceedings of 11th International Conference on Principles of Knowledge Representation and Reasoning* (Menlo Park, CA: AAAI Press, 2008), 378–389.

Chapter 8

1. J. Pearl, *Causality* (Cambridge: Cambridge University Press, 2009).

Chapter 9

1. M. Kearns and L. G. Valiant, "Cryptographic Limitations on Learning Boolean Formulae and Finite Automata," *Journal of the ACM* 41, no. 1 (1994): 67–95. Preliminary version in *Proceedings of the 21st ACM Symposium on Theory of Computing* (1989): 433–444.

2. R. Schapire, "Strength of Weak Learnability," *Machine Learning* 5 (1990): 197–227.

3. Y. Freund and R. E. Schapire, "A Decision-Theoretic Generalization of On-Line Learning and an Application to Boosting," *Journal of Computer and System Sciences* 55, no. 1 (1997): 119–139.

4. M. Aizerman, E. Braverman, and L. Rozonoer, "Theoretical Foundations of the Potential Function Method in Pattern Recognition Learning," *Automation and Remote Control* 25 (1964): 821–837.

5. B. E. Boser, I. M. Guyon, and V. N. Vapnik, "A Training Algorithm for Optimal Margin Classifiers," *5th Annual ACM Workshop on Computational Learning*

Theory (Pittsburgh, PA: ACM Press, 1992), 144–152; C. Cortes and V. Vapnik, "Support-Vector Networks," *Machine Learning* 20 (1995).

6. T. M. Mitchell et al., "Learning to Decode Cognitive States from Brain Images," *Machine Learning* 57 (2004): 145–175.

7. A. M. Turing, "Intelligent Machinery" (unpublished manuscript, 1948). Reproduced in B. J. Copeland, *The Essential Turing* (Oxford: Oxford University Press, 2004), 410–432.

8. L. G. Valiant, "Functionality in Neural Nets," *Proceedings of the First Workshop on Computational Learning Theory* (San Francisco: Morgan Kaufmann Publishers, 1988), 28–39.

9. R. Paturi, S. Rajasekaran, and J. H. Reif, "The Light Bulb Problem," *Proceedings of the Second Annual Workshop on Computational Learning Theory* (San Francisco: Morgan Kaufmann Publishers, 1989), 261–268; P. Indyk and R. Motwani, "Approximate Nearest Neighbors: Towards Removing the Curse of Dimensionality," *Proceedings of the 30th Annual ACM Symposium on Theory of Computing* (1998): 604–614; M. Dubiner, "Bucketing, Coding and Information Theory for the Statistical High Dimensional Nearest Neighbor Problem," arXiv:0810.4182 (2008); G. Valiant, "Finding Correlations in Subquadratic Time, with Applications to Learning Parities and Juntas," *53rd Annual IEEE Symposium on Foundations of Computer Science* (2012): 11–20.

10. This is discussed more fully in L. G. Valiant, *Circuits of the Mind* (New York: Oxford University Press, 1994, 2000).

11. H. Gardner, *Frames of Mind: The Theory of Multiple Intelligences* (New York: Basic Books, 1983).

12. A. M. Turing, "Computing Machinery and Intelligence," *Mind* 49 (1950): 433–460. Notwithstanding the title, Turing does not claim in the text a definition of intelligence, but rather a criterion on when a machine could be regarded as "thinking." However, from the beginning, the Turing Test has been regarded as a fitting reference point for any discussion of intelligence.

13. Ibid.

Chapter 10

1. Winston Churchill, Speech at Harvard University, September 6, 1943.

Glossary

An **accessible target** is a function that can be learned or evolved because it is within a learnable or evolvable class with respect to the available features.

A **Boolean function** is a function whose inputs and outputs take just true or false values. These values may be represented by 1 and 0, or by 1 and –1. An example of a Boolean function of two arguments is the **or** function, written as **or**(x, y) and defined to have value true if and only if at least one of the two arguments x, y has value true.

BPP (bounded probabilistic polynomial) computations are those that can be performed in polynomial time by a randomized Turing machine.

BQP (bounded quantum polynomial) computations are those that can be performed in polynomial time by a quantum Turing machine.

A **circuit** is a computation where the dependencies among various input, output, or intermediate values can be made explicit.

A **complexity class** is a set of problems characterized by the computations that can solve them. For example, P and NP are complexity classes.

A function is **computable** if it can be computed by some Turing machine, in the sense that for every input the Turing machine produces the answer within some finite number of steps.

Concept is a term used here to denote a function in the context of learning.

A **conjunction** is a Boolean function that has value true if all of its arguments have value true.

A **disjunction** is a Boolean function that has value true if at least one of its arguments has value true.

Ecorithm is a term introduced here to denote an algorithm that takes information from its environment so as to perform better in that environment. Algorithms for machine learning, evolution, and for learning for the purpose of reasoning are all instances of ecorithms.

A class of functions is **evolvable** if there is an evolution algorithm that can evolve every member of it using only polynomial resources and achieving polynomial error control.

The **expression level** of a protein in a cell is a measure of how much of the protein is being produced.

Feasible computation is identified here with computations in which the number of steps is bounded by a polynomial in terms of the number of bits required to write down the input.

A **function** is a mathematical assertion of a specific dependence of a value on some variables, parameters, conditions, or arguments. For example, the function $f(x, y) = 2x + 3y$ is the dependence that f is the sum of twice the first argument and thrice the second.

The **ideal function** specifies for a particular evolving entity in a particular environment the best possible action for every possible combination of conditions.

The **input function** of a protein is the function that determines the expression level of the protein in terms of all the relevant conditions in the cell.

Intelligence is generally used in the text in the colloquial sense of human intelligence, but the aspect of it that is addressed more technically is that of reasoning on uncertain, learned knowledge.

A **linear inequality** is an assertion that the value of some linear combination of variables is greater or less than some value (e.g., $3x + 6y - 8z < 7$).

A **linear separator** for a set of labeled examples is a linear inequality that satisfies all the positive examples and none of the negative examples.

Nondeterministic computations are those that perform an exponential search for a solution in parallel.

NP (nondeterministic polynomial) is the class of problems for which there are nondeterministic computations where each of the parallel branches of the search uses at most a polynomial number of steps in terms of the number of bits of the input.

A problem is **NP-complete** if its polynomial time solution would imply polynomial time solutions for every problem in NP.

P is the class of problems to which solutions can be found by deterministic computations taking a polynomial number of steps in terms of the number of bits of the input.

#P ("sharp" P) is the class of problems that count the number of solutions found in an NP computation.

PAC (probably approximately correct) learning is the process of learning from examples, where the number of computational steps is polynomially bounded and the errors are polynomially controlled.

A class of problems is **PAC learnable** if there is a learning algorithm that can learn every member of the class, using only polynomial resources and achieving polynomial error control.

PAC semantics is the sense in which the definition of PAC learning guarantees accuracy.

A **parity function** is a Boolean function that has value true if and only if an odd number of its arguments have value true.

The **Perceptron** algorithm is a specific method for learning linear inequalities.

Polynomially bounded for a function $f(n)$ is used here to mean that for some fixed numbers c and k, for every positive integer value of n, $f(n) < cn^k$.

A computational **problem** is a function that is to be evaluated. For example, determining how many factors a number has is a computational problem. A **solution** to such a problem is an algorithm that evaluates that function.

The **protein expression network** represents how the expression levels of all the proteins in a cell are regulated in terms of each other and other relevant factors. It is sometimes referred to as the **gene expression network.**

A **randomized Turing machine** is a Turing machine that at any step may choose one among a set of possible transitions by making a random decision according to the toss of a coin.

Resilience to different distributions is the desirable property of a learning algorithm to give reliable answers for wide ranges of distributions of the examples.

Resilience to noise is the desirable property of a learning algorithm to give answers that are degraded only a little by any noise in the data from which it is learning.

A robust computational model for a phenomenon is one that is provably equivalent to a wide range of alternative definitions of computational models for that phenomenon. Turing machines for the phenomenon of computation offer the exemplary paradigm.

Robust logic is a system in which learning and reasoning have a common semantics and both can be accomplished feasibly in the PAC sense.

Robustness to computation and data is a way of phrasing what PAC learning accomplishes, namely the requirement that it should be practicable to drive down errors arbitrarily by increasing the amounts of training data and computation appropriately.

A **statistical query** algorithm is a learning algorithm that can receive information about examples only by asking statistical questions about them, rather than by processing individual examples.

Target pursuit is the capability in both learning and evolution to pursue a large number of accessible targets simultaneously.

Theoryful is a term defined here to denote decisions for which there is a good explanatory and predictive theory, such as a scientific theory.

Theoryless is a term defined here to denote decisions that are not known to be theoryful.

A **Turing machine** is a model of computation that is widely believed to encompass all information processing that one would think of as mechanistic.

Acknowledgments

As the text makes clear, this book is deeply rooted in the visionary ideas of Alan Turing. The synthesis offered here is within the framework of computational learning theory. Over the last three decades many have enriched this field, and I would particularly like to thank Dana Angluin, Avrim Blum, Andrzej Ehrenfeucht, Vitaly Feldman, Yoav Freund, David Haussler, Varun Kanade, Michael Kearns, Roni Khardon, Adam Klivans, Ming Li, Nick Littlestone, Yishay Mansour, Loizos Michael, Lenny Pitt, Ron Rivest, Dan Roth, Robert Schapire, Rocco Servedio, and Manfred Warmuth. It is also a pleasure to acknowledge the early pioneers of computational complexity, including Manuel Blum, Stephen Cook, Juris Hartmanis, Richard Karp, Michael Rabin, and Volker Strassen.

I am grateful to Juliet Harman for her careful and critical reading of the manuscript and for her many ideas that have improved this book.

I would also like to thank Thomas Kelleher at Basic Books for his valuable editorial suggestions.

It is traditional for authors to thank their spouse for patiently suffering the inevitable hardships as a writer's companion. I can only report unreserved enthusiasm for the project, and thank Gayle for that and for her significant and extensive suggestions on the text.

Index

n means note, g means glossary